Notes on the Witt Classification of Hermitian Innerproduct Spaces over a Ring of Algebraic Integers

Notes on the Witt Classification of Hermitian Innerproduct Spaces over a Ring of Algebraic Integers

by P. E. Conner
Nicholson Professor of Mathematics
Louisiana State University, Baton Rouge

University of Texas Press, Austin

For reasons of economy and speed this volume has been printed from camera-ready copy furnished by the author, who assumes full responsibility for its contents.

International Standard Book Number
ISBN: 978-0-292-74067-9
Library of Congress Catalog Card Number
79-87884

Printed in the United States of America
First paperback printing, 2012

Contents

Introduction

These notes are intended to be an introduction to the Witt classification of Hermitian innerproducts as a preparation for the study of applications of this subject to topological problems in knot concordance, cobordism classification of diffeomorphisms and actions of finite groups. At the same time we hope to acquaint the algebraic number theorist with the questions about innerproduct spaces which confront the topologist. We should also mention that at least for cyclic groups of prime power order, C_{p^r}, there appears to be an immediate relationship between the material discussed in these notes and the Wall surgery obstruction groups $L_0^h(C_{p^r})$ and $L_0^p(C_{p^r})$.

We shall presume a background of familiarity with algebraic number theory equivalent to a year's course on that topic. Our refereneces will provide all the basic ideas we use. At one point we mention the Hilbert Class Field and the Artin Reciprocity Map. The reference to the class field is not essential and the Artin map is used only for an unramified quadratic extension in which case we can point out the properties of the Artin map as being derived from Hilbert symbol reciprocity.

Our viewpoint has been to develop an exposition of the Witt ring of Hermitian innerproducts over the ring of integers in a quadratic extension of an algebraic number field in a manner which will allow comparison to, and contrast with, the exposition of the Witt ring of symmetric innerproducts over the ring of algebraic integers found in [M-H]. The emphasis on Witt classification reflects the fact that this is the álgebraic analogue of cobrodism classification in topology. We are motivated to take up the Hermitian case because it is our observation that this case dominates applications to topological problems. There is not surprisingly a large literature concerning the Witt classification of Hermitian innerproducts, together with the generalizations of the notion of Hermitian, over quite general rings.

We have focused on the classical case first because it as such definitely relates to cobordism classification questions and second because by appeal to algebraic number theoretical techniques it is possible to obtain explicit computations demonstrating what can happen in this context of Hermitian Witt. In many applications it is useful, and sometimes even necessary, to replace the ring of integers by an overring [ACH] or by a non-maximal order [St]. But what we discuss here must be done first and thus we hope to have provided the basis from which the reader can make the appropriate modifications as the need arises.

Our first chapter is devoted to the basic number theory required. For relative quadratic extensions we first point out how primes, both finite and infinite behave. Then we introduce the formalism of Hilbert symbols which occupy the most central position in these notes. In two sections we exhibit relations, recognized from the foundations of algebraic number theory, between ideal class groups and units. We worked out the exposition here to provide our reader an opportunity to develop a feeling of coexistence with class groups. These matters should not be ignored. To do so will surely lead to errors in later computations. In a separate section we take up unramified quadratic extensions (no prime, finite or infinite, ramifies). We close off the chapter by discussion examples to illustrate our points.

In Chapter II we do $H(E)$, the Witt ring of Hermitian innerproducts over a field E with an automorphism of period 2. We give arguments in detail, generally parallel to those in [M-H]. Our proofs are framed to generalize quickly to the structures occurring in topological applications. The chapter ends with the determination of $H(E)$ from Landherr's paper [La].

The third chapter discusses torsion innerproducts at a level of generality suitable for our objectives. We work out the structure of finitely generated torsion modules over a Dedekind domain although this may be found in textbooks on commutative rings. We need a specific way of doing it and we wish to point out clearly the relation of the torsion innerproduct to the module structure. The chapter ends with the determination of the Witt groups of the torsion innerproducts.

Chapter IV takes up $H_u(K)$. Here $u \in O_E^*$ is a unit for which $u\bar{u} = 1$ and $K \subset E$ is a fractional ideal for which $K = \bar{K}$. We proceed to consider pairs (P, h) with P a finitely generated projective O_E-module and $h: P \times P \to K$ a biadditive function which satisfies

$$\lambda h(v,w) = h(\lambda v,w) = h(v,\bar{\lambda}w)$$

for every algebraic integer $\lambda \in O_E$ and for which

$$\overline{uh(v,w)} = h(w,v)$$
$$adj_h: P \simeq Hom_{O_E}(P,K) .$$

A Witt classification is introduced and the group $H_u(K)$ is studied and eventually computed. If $u = 1$, $K = O_E$ then the ring $H(O_E)$ results and we show that each $H_u(K)$ is an $H(O_E)$-module which naturally embeds into $H(O_E)$ as an ideal. Let us explain our use of the $H_u(K)$ device. Our first experience, [ACH], was with $H(O_E)$ and $H_{-1}(O_E)$, the latter regarded as skew-Hermitian innerproducts. However Stoltzfus [St] points out to us that the restriction to $H_{\pm 1}(O_E)$ was unsatisfactory for the application he wished to make to knot concordance. He found that, with $\mathcal{D}^{-1}$ the inverse different ideal of E/Q, it was the Witt groups $H_{\pm 1}(\mathcal{D}^{-1})$ he needed. Later Warschauer found that $H_u(\mathcal{D}^{-1})$ with $u\bar{u} = 1$ was necessary to his research on the Witt classification of innerproduct spaces over the rational integers about which no assumption of symmetry is made. (We note parenthetically that this asymmetric Witt group has begun to appear in topology, for example in Quinn's research on the relative open book problem.) Our problem was to reconcile and unify these approaches which we did by using $H_u(K)$.

The last chapter is a review of the Witt ring of symmetric innerproduct spaces over the ring of algebraic integers. The purpose is simply to allow comparison with the Hermitian case. It will be seen that there are marked differences.

We think of the ideal class group as playing the role of the reduced projective class group, and indeed it is isomorphic to $\tilde{K}_0(O_E)$. And in similar vein we think of the group of units O_E^* as analogous to $K_1(\cdot)$. With this viewpoint (I.4.4) is then a parallel of the Rothenberg exact sequence. For example we speak of a unit which is a norm from the field but not the norm of a unit. This translates into a Hermitian innerproduct space structure on a free O_E-module of rank 2 which is Witt trivial but for which there is no metabolizer which is also free. This is the same phenomenon which for a cyclic group of odd prime power order accounts for a kernel in the forgetful homomorphism $L_0^h(C_{p^r}) \to L_0^p(C_{p^r})$. Again a cokernel for the homomorphism $H^2(C_2;O_E^*) \to \text{Gen}(E/F)$ will mean that we can find Witt classes in $H(O_E)$ which cannot be represented by a Hermitian innerproduct on a free O_E-module.

The reader should be warned explicitly about one point in particular. We are familiar with the process by which a stable determinant, a unit in the (commutative) ground ring, can be assigned to an automorphism of a finitely generated projective module. However this does not work for an innerproduct space structure. We shall see easily constructed examples of Hermitian innerproduct spaces for which the determinant, even modulo norms, is not a unit. We ask our reader to remain aware of this.

We should mention some of our usages. We use the terms finite prime and infinite prime. For finite primes we refer to the order of an element or of a fractional ideal at that prime. Obviously 0 always has order $+\infty$. Real infinite primes are thought of as orderings of the algebraic number field. We use the expression localized completion rather than the more usual term completion. This is because for a finite prime we think of a two step process. First the ring of integers is localized at a prime and then this local ring is completed with respect to the powers of the maximal ideal and finally the field of fractions is the local-zed completion. The Galois group here will always be the cyclic group of order 2, C_2. If an abelian group A is equipped with an automorphism of period 2, $a \to \bar{a}$, then $H^2(C_2;A)$ is the quotient of the group of elements for which $a = \bar{a}$ by the subgroup of elements of the form $b\bar{b}$. Similarly $H^1(C_2;A)$ is the group of elements for which $a\bar{a} = 1$ modulo the subgroup of elements of the form $b\bar{b}^{-1}$. To the question as to why we use this cohomology framework we can only reply it is because to us that is what it looks like it should be. For examples we have made it a point to use quadratic extensions of Q whenever possible. If this fails then we turn to biquadratic extensions of Q. There are ample illustrations involving cyclotomic number fields in [ACH] and examples involving practically everything in [St].

We regard the notes as a graduate level exposition. Accordingly we have held our references down to a core of three, [L],[O'M] and [M-H]; texts which we think should be available to every graduate student. The reference [C] is concerned with quadratic extensions of Q. The references [ACH], [M-H] and [St] will provide a good list of recent publications on Witt classification and its relation to topological questions.

The present notes arose from the collaboration in writing [ACH] as well as from many conversations with Neal Stoltzfus about his paper [St]. We were also influenced by Max Warschauer's thesis written at LSU. We extend our appreciation to Hubert Butts, Craig Cordes, Gordon Paul and Luther Wade, the algebraic number theorists at LSU who gave us their time in an effort which enabled us to develop the algebraic background needed for these notes.

The notes were written during a sabbatical leave from LSU for the spring semester, 1978. Our thanks are expressed to the Mathematics Department of the University of Texas, Austin for an invitation to a "mini-professorship" which gave us cause to pull this all together and to try it out. We are particularly grateful to Nita Goldrick of U.T. Austin for the typing of our manuscript and to the University of Texas Press for publishing these notes.

Baton Rouge, LA.
February, 1978

Notes on the Witt Classification of Hermitian Innerproduct Spaces over a Ring of Algebraic Integers

I Relative Quadratic Extensions

1. Extension of primes

By E/F we shall denote an extension, E, with degree 2, of the algebraic number field F. We denote by $x \to \bar{x}$ the associated involution of E with fixed field F. Let O_E and O_F be the rings of algebraic integers in the respective fields E and F.

We shall use the symbols $\mathcal{P}$ and P to denote prime ideals in O_E and O_F. These are to be called <u>finite primes</u>. An infinite prime will be denoted by $\mathcal{P}_\infty$ or P_∞. In particular we always regard a real infinite prime as an ordering of the algebraic number field.

For the finite primes P in O_F we shall distinguish three cases:

1. <u>Split</u>: $P \cdot O_E = \mathcal{P}\bar{\mathcal{P}}$ for $\mathcal{P}$ a prime in O_E with $\mathcal{P} \neq \bar{\mathcal{P}}$.
2. <u>Inert</u>: $P \cdot O_E = \mathcal{P}$ a prime in O_E (for which $\mathcal{P} = \bar{\mathcal{P}}$).
3. <u>Ramified</u> $P \cdot O_E = \mathcal{P}^2$, $\mathcal{P}$ a prime in O_E (for which $\mathcal{P} = \bar{\mathcal{P}}$ also).

In F^* there is an element σ, unique up to multiplication by a non-zero square in F, for which $E = F(\sqrt{\sigma})$. If P_∞ is a real infinite prime in F then P_∞ is <u>ramified</u> if and only if $\sigma < 0$ in that ordering. If $\sigma > 0$ then the real infinite prime is split. All complex infinite primes in F are regarded as split and generally ignored.

Note that for any ramified real infinite prime in F every norm from E^*, $x^2 - y^2\sigma$, is positive in that ordering. Furthermore, with $\sqrt{\sigma}$ chosen, we can associate uniquely to each ramified real infinite prime an embedding of E into $\mathbb{C}$, equivariant with respect to complex conjugation and sending $\sqrt{\sigma}$ to a point on the positive imaginary axis. This embedding extends the embedding of F into

$\mathbb{R}$ associated with the ramified real infinite prime and is the complex prime of E into which it ramifies. On the other hand, if the real infinite prime of F splits then the ordering extends to two distinct orderings of E which are conjugate in the sense that if E^+ is the set of positive elements in one of the two extensions then $\overline{E}^+$ is the set of positive elements in the other.

Let us present a lemma about finite primes for the reader's verification

(1.1) Lemma: *If* $y \in F^*$ *and if*

a) $\mathcal{P} \neq \overline{\mathcal{P}}$ *is over* P *split then*

$$\text{ord}_P\, y = \text{ord}_{\mathcal{P}}\, y = \text{ord}_{\overline{\mathcal{P}}}\, y\ .$$

b) $\mathcal{P} = \overline{\mathcal{P}}$ *is over inert* P *then*

$$\text{ord}_P\, y = \text{ord}_{\mathcal{P}}\, y\ .$$

c) $\mathcal{P} = \overline{\mathcal{P}}$ *is over ramified* P *then*

$$2\ \text{ord}_P\, y = \text{ord}_{\mathcal{P}}\, y\ . \quad \blacksquare$$

Here *order* refers to the exponent with which P or $\mathcal{P}$ appears in the factorization of the principal ideal (fractional principal ideal) $y\,O_F$ or $y\,O_E$. We shall also use ν_P and $\nu_{\mathcal{P}}$ to denote order. The order of zero is $+\infty$.

Now let $O_F(P)$, $O_E(\mathcal{P})$ be the rings of local integers associated with the finite primes. A *local uniformizer* is an element of order 1 at the prime. If $P \to \mathcal{P}\overline{\mathcal{P}}$ is split then a local uniformizer $\pi \in O_F(P)$ is a local uniformizer for $O_E(\mathcal{P})$ and $O_E(\overline{\mathcal{P}})$. If $P \to \mathcal{P}$ is inert then a local uniformizer $\pi \in O_F(P)$ is also a local uniformizer for $O_E(\mathcal{P})$. If $P \to \mathcal{P}^2$ is ramified then for any local uniformizer $\pi \in O_E(\mathcal{P})$ the norm $\pi\overline{\pi}$ is a local uniformizer for $O_F(P)$.

If $\mathcal{P} = \overline{\mathcal{P}}$ then the involution makes the group of local units into a C_2-module. Thus $H^1(C_2;O_E(\mathcal{P})^*)$ and $H^2(C_2;O_E(\mathcal{P})^*)$ are defined.

(1.2) Lemma: If $\mathcal{P} = \overline{\mathcal{P}}$ is an involution invariant prime ideal of O_E then

$$H^1(C_2;O_E(\mathcal{P})^*) = \{1\}, \quad \mathcal{P} \text{ over inert}$$

$$H^1(C_2;O_E(\mathcal{P})^*) \approx C_2, \quad \mathcal{P} \text{ over ramified.}$$

Proof:

If $u \in O_E(\mathcal{P})^*$ is a local unit for which $u\bar{u} = 1$ then by Hilbert Theorem 90 there is a $z \in E^*$ with $z\bar{z}^{-1} = u$. We can write $z = \pi^{\nu_{\mathcal{P}}(z)} v$ for some v in $O_E(\mathcal{P})^*$. If $\mathcal{P}$ is over inert we may assume $\pi = \bar{\pi}$ so $z\bar{z}^{-1} = v\bar{v}^{-1} = u$ and hence $H^1(C_2;O_E(\mathcal{P})^*)$ is trivial in that case.

If $\mathcal{P}$ is over ramified we note $\pi\bar{\pi}^{-1} \in O_E(\mathcal{P})^*$ so that

$$u = (\pi\bar{\pi}^{-1})^{\nu_{\mathcal{P}}(z)} v\bar{v}^{-1}$$

and hence $c\ell(u) = (c\ell(\pi\bar{\pi}^{-1}))^{\nu_{\mathcal{P}}(z)}$ in $H^1(C_2;O_E(\mathcal{P})^*)$. This shows that $H^1(C_2;O_E(\mathcal{P})^*)$ is generated by $c\ell(\pi\bar{\pi}^{-1})$. This class is non-trivial for if $\pi\bar{\pi}^{-1} = v\bar{v}^{-1}$ for some $v \in O_E(\mathcal{P})^*$ then $\overline{\pi\bar{v}} = \bar{\pi}v = \pi\bar{v} \in F^*$. This is a contradiction since $2 \operatorname{ord}_P \pi\bar{v} = \operatorname{ord}_{\mathcal{P}} \pi\bar{v} = 1$. Of course $c\ell(\pi\bar{\pi}^{-1})$ is independent of the choice of local uniformizer. ■

If $O_E(\mathcal{P})$ is replaced by $\tilde{O}_E(\mathcal{P})$, the ring of integers in the localized completion of E at $\mathcal{P}$, then we may state

(1.3) Lemma: If $\mathcal{P} = \overline{\mathcal{P}}$ over inert, then $H^2(C_2;\tilde{O}_E(\mathcal{P})^*) = \{1\}$ while if $\mathcal{P} = \overline{\mathcal{P}}$ over ramified then $H^2(C_2;\tilde{O}_E(\mathcal{P})^*) \simeq C_2$.

Proof:

This is a consequence of the local norm index theorem [L, p. 188]. If P is inert then every unit in $\tilde{O}_P(P)^*$ is the norm of a unit in $\tilde{O}_E(\mathcal{P})^*$, while if P is ramified the norms from $\tilde{O}_E(\mathcal{P})^*$ form a subgroup of index 2 in $\tilde{O}_F(P)^*$. Incidentally this is a good place for the reader to convince himself that if $u \in \tilde{O}_F(P)^*$ is a norm from the localized completion of E at $\mathcal{P}$ then it is the norm of a unit in $\tilde{O}_E(P)^*$. As we shall see, the global version of this observation is not generally valid. ■

At this point we need a (preliminary) division of the finite ramified primes into two types.

(1.4) Definition: If $P \subset O_F$ is ramified then P is of type I if and only if $c\ell(-1) = c\ell(\pi\bar{\pi}^{-1}) \in H^1(C_2; O_E(\mathcal{P})^*)$. If $c\ell(-1) = 1$ in $H^1(C_2; O_E(\mathcal{P})^*)$ then P is of type II.

Let us immediately go into a lemma.

(1.5) Lemma: The ramified prime is of type I if and only if there is a local uniformizer $\pi \in O_E(\mathcal{P})$ for which $\bar{\pi} = -\pi$. The ramified prime is of type II if and only if there is a local unit $v \in O_E(\mathcal{P})^*$ for which $\bar{v} = -v$.

Proof:

By definition $c\ell(-1) = 1$ if and only if $v\bar{v}^{-1} = -1$ for some local unit. This takes care of type II. Then P is of type I if and only if $c\ell(-1) = c\ell(\pi\bar{\pi}^{-1})$ for every local uniformizer $\pi \in O_E(\mathcal{P})$. That is if and only if $-1 = \pi\bar{\pi}^{-1} v\bar{v}^{-1}$ for some $v \in O_E(\mathcal{P})^*$. Obviously $\overline{\pi v} = -\pi v$ so πv is the skew local uniformizer. ■

Let $m(P) \subset O_F(P)$, $m(\mathcal{P}) \subset O_E(\mathcal{P})$ be the unique maximal ideals. Then $F_P = O_F(P)/m(P)$ and $E_{\mathcal{P}} = O_E(\mathcal{P})/m(\mathcal{P})$ are residue fields. A finite prime is dyadic or non-dyadic according to whether or not the residue field has characteristic 2.

(1.6) Lemma: If P is a non-dyadic ramified prime then P is of type I.

Proof:

If $v\bar{v}^{-1} = -1$ for some $v \in O_E(\mathscr{P})^*$ then because the induced involution on the residue field is the identity we would find that $1 = -1$ in $E_{\mathscr{P}}$. This is impossible since $E_{\mathscr{P}} = F_P$ has odd characteristic. ■

By contrast a dyadic ramified prime may be of either type. Using (I.1.5) it follows that 2 is of type II in $Q(\sqrt{-1})/Q$ while 2 is of type I in $Q(\sqrt{2})/Q$. What type is 2 in $Q(\sqrt{3})/Q$?

Clearly we shall conclude by characterizing type by the parity of the local differential exponent. Suppose P ramifies into $\mathscr{P}^2$, then there is always an element $\rho \in \tilde{O}_E(\mathscr{P})$ such that $1, \rho$ is a basis for $\tilde{O}_E(\mathscr{P})$ as an $\tilde{O}_F(P)$ module [L, p. 59]. Then ρ is a root of $\rho\bar{\rho} - (\rho + \bar{\rho})t + t^2$, and $\rho - \bar{\rho}$ is the different of the local extension $\tilde{E}(\mathscr{P})/\tilde{F}(P)$. Write $\rho - \bar{\rho} = \pi^d v$ for π some local uniformizer at $\mathscr{P}$ and $v \in \tilde{O}_E(\mathscr{P})^*$. The integer d is the differential exponent at P. Now $\rho - \bar{\rho}$ is skew so

$$-\pi^d v = (\bar{\pi})^d \bar{v}$$

or

$$-v\bar{v}^{-1} = (\bar{\pi}\pi^{-1})^d$$

which shows $c\ell(-1) = c\ell(\pi\bar{\pi}^{-1})^d$. We can put our answer in terms of the different ideal $\mathfrak{D}_{E/F} \subset O_E$ since $\text{ord}_{\mathscr{P}} \mathfrak{D}_{E/F} = d$ for $\mathscr{P} = \bar{\mathscr{P}}$ over ramified [L, p. 61].

(1.7) Lemma: If $\mathfrak{D}_{E/F} \subset O_E$ is the different ideal of the extension E/F then $\text{ord}_{\mathscr{P}} \mathfrak{D}_{E/F} \equiv 1 \pmod 2$ at all $\mathscr{P} = \bar{\mathscr{P}}$ over non-dyadic ramified. If $\mathscr{P} = \bar{\mathscr{P}}$ is over dyadic ramified then $\text{ord}_{\mathscr{P}} \mathfrak{D}_{E/F} \equiv 1$ or $0 \pmod 2$ according to whether P is of type I or II. ■

2. Hilbert symbols

In this section we are concerned with $H^2(C_2;E^*)$; that is, with the quotient of F^* by the subgroup of norms from E^*. The standard calculations are based on a special case of the Hasse Cyclic Norm Theorem [O'M, p. 186]. If $\mathcal{P}$ is a prime in E, finite or infinite, then $\mathcal{P}$ lies over (is an extension of) a unique prime P in F. There results a (local) extension of the respective localized completions. The local extension is of degree 1 if and only if P is split. This means that σ already has a square root in the completion $\tilde{F}(P)$. Otherwise the local extension is of degree 2.

(2.1) Hasse: An element $y \in F^*$ is a norm from E^* if and only if $y \in \tilde{F}(P)^*$ is a norm from $\tilde{E}(\mathcal{P})^*$ for every prime, finite or infinite in E. ■

Obviously split primes may be ignored since $\tilde{E}(\mathcal{P}) = \tilde{F}(P)$. What is needed then is an invariant which recognizes for inert or ramified primes whether or not y is a local norm. From the local norm index theorem we also expect this will take the form of a mod 2 invariant. The Hilbert symbol (norm residue symbol) with values in Z^* will do just this. We refer to [O'M, p. 164] for the definition and properties of Hilbert symbols. We choose a $\sigma \in F^*$ with $E = F(\sqrt{\sigma})$. Then for each P in F, and $y \in F^*$ there is the Hilbert symbol $(y,\sigma)_P \in Z^*$. This symbol is 1 at all split primes, it is 1 for almost all primes and Hilbert's Reciprocity Theorem asserts

$$\prod_P (y,\sigma)_P = 1,$$

where product is over all primes in F. We shall state as a corollary of (I.2.1)

(2.2) Theorem: An element $y \in F^*$ is a norm if and only if $(y,\sigma)_P = 1$ for all primes in F. ■

We may think of $y \to (y,\sigma)_P$ as defining a family of homomorphisms h_P of $H^2(C_2;E^*)$ into Z^*. They are independent of the choice of σ with $E = F(\sqrt{\sigma})$. The properties of h_P are:

a) $h_P(c\ell(y)) = 1$ for almost all P

b) $h_P(c\ell(y)) = 1$ if P splits

c) $\Pi\, h_P(c\ell(y)) = 1$

d) $h_P(c\ell(y)) = 1$ for all P if and only if $c\ell(y) = 1$

e) if to each P in F there is assigned $\varepsilon_P \in Z^*$ so that $\varepsilon_P = 1$ for almost all P, $\varepsilon_P = 1$ for all P split and $\Pi\, \varepsilon_P = 1$ then there is a unique $c\ell(y) \in H^2(C_2;E^*)$ with $h_P(c\ell(y)) = \varepsilon_P$ for all P.

Part e) is a consequence of the realization of Hilbert symbols [O'M, p. 203]. We shall refer to this frequently.

We have seen that the Hilbert symbol is trivial at split primes. Then let P be inert, so that if π is a local uniformizer of $O_F(P)$ it is also a local uniformizer for $O_E(\mathcal{P})$. Thus $(\pi,\sigma)_P = -1$; that is, π cannot be a local norm because $\mathrm{ord}_P\, \pi = 1 = \mathrm{ord}_{\mathcal{P}}\, \pi$. For $y \in F^*$ write $y = \pi^{\nu_P(y)} v$, $v \in O_F(P)^*$. At an inert prime every local unit is a local norm, so using the bimultiplicative property of the symbols

$$(y,\sigma)_P = ((\pi,\sigma)_P)^{\nu_P(y)} (v,\sigma)_P = (-1)^{\nu_P(y)} .$$

We may state

(2.3) <u>Lemma</u>: <u>If</u> P <u>is inert then</u>

$$(y,\sigma)_P = (-1)^{\nu_P(y)} = h_P(c\ell(y))$$

<u>for all</u> $y \in F^*$. ■

For a finite ramified prime the situation is as follows. Choose a local uniformizer π in $O_E(\mathscr{P})$, then the norm $\pi\bar{\pi}$ is a local uniformizer for $O_F(P)$. Hence for $y \in F^*$ write $y = (\pi\bar{\pi})^{\nu_P(y)} U$, $U \in O_F(P)^*$. Then

$$(y,\sigma)_P = (U,\sigma)_P.$$

We can say, [O'M, p. 166],

(2.4) Lemma: <u>If P is a non-dyadic ramified prime and $U \in O_P(F)^*$ is a local unit then $(U,\sigma)_P = 1$ if and only if U is a square in the residue field.</u>

<u>Proof</u>:

Assume first U is a square in F_P. Because F_P has odd characteristic there will be two distinct square roots. By Hensel's Lemma then $t^2 - U$ will factor in the completion $\tilde{F}(P)$, which is the statement that U is a square in $\tilde{O}_F(P)^{**}$.

If U is a local norm from $\tilde{E}(\mathscr{P})^*$ then $U = V\bar{V}$ for some $V \in \tilde{O}_E(\mathscr{P})^*$. But then in the residue field $\tilde{O}_E(\mathscr{P})/\tilde{m}(\mathscr{P}) = E_{\mathscr{P}} = F_P$ we have $U = V^2$. ■

To close the section we point out

(2.5) Lemma: <u>If P_∞ is a ramified real infinite prime then $(y,\sigma)_{P_\infty} = 1$ if and only if $y > 0$ for P_∞.</u> ■

3. The group Gen(E/F)

In this section we introduce a group Gen(E/F) which eventually will be embedded into the group of units of the Witt ring $H(O_E)$ multiplicatively. Furthermore, Gen(E/F) will receive the discriminant from $H(O_E)$. Our viewpoint was based on the need to get an enlargement of the group of units O_F^* which does not in fact have a satisfactory relation to $H(O_E)$.

Begin with the group of all pairs (y,A) in which $y \in F^*$ and $A \subset E$ is a fractional O_E-ideal for which $yA\bar{A} = O_E$. The pairs form a group with multiplication

$$(y,A)(y_1,A_1) = (yy_1, AA_1).$$

The identity is $(1,O_E)$ and (y^{-1},A^{-1}) is the inverse of (y,A). There is an obvious embedding of O_F^* into this group of pairs.

To obtain $\mathrm{Gen}(E/F)$ we factor out the subgroup of all pairs of the form $(z\bar{z}, z^{-1}B\bar{B}^{-1})$ where $z \in E^*$ and B is a fractional O_E-ideal. The resulting quotient group is $\mathrm{Gen}(E/F)$ and an element in this group is denoted by $\langle y,A\rangle$.

(3.1) Lemma: <u>If $y \in F^*$ then there is a fractional O_E-ideal A for which $yA\bar{A} = O_E$ if and only if $\mathrm{ord}_P\, y \equiv 0 \pmod 2$ for all P inert; that is, if and only if $(y,\sigma)_P = 1$ for all inert P.</u>

<u>Proof</u>:

For any $y \in F^*$ we know

$$\mathrm{ord}_{\mathcal{P}}\, y = \mathrm{ord}_{\bar{\mathcal{P}}}\, y = \mathrm{ord}_P\, y, \quad P \text{ split}$$

$$\mathrm{ord}_{\mathcal{P}}\, y = 2\, \mathrm{ord}_P\, y, \quad P \text{ ramified}$$

so that if we add the hypothesis $\mathrm{ord}_P\, y \equiv 0 \pmod 2$ for all inert we can make up a fractional O_E-ideal A with $\mathrm{ord}_{\mathcal{P}}\, y + \mathrm{ord}_{\mathcal{P}}\, A + \mathrm{ord}_{\mathcal{P}}\, \bar{A} = \mathrm{ord}_{\mathcal{P}}\, y + \mathrm{ord}_{\mathcal{P}}\, A + \mathrm{ord}_{\bar{\mathcal{P}}}\, A = 0$ for every finite $\mathcal{P}$. Surely $yA\bar{A} = O_E$. The converse is obvious. ■

We want to calculate $\mathrm{Gen}(E/F)$. First we turn to $I(E)$ the group of all fractional O_E-ideals. This is a C_2-module by $A \to \bar{A}$. Now $I(E)$ is a free abelian group generated by (the) prime ideals in O_E and C_2 acts by permuting these generators. Thus if $\mathcal{F} \subset I(E)$ is the subgroup generated by those primes for which $\mathcal{P} = \bar{\mathcal{P}}$, then it is an elementary exercise to show

(3.2) Lemma: For the C_2-module $I(E)$

$$H^2(C_2;I(E)) \simeq \mathcal{F}/\mathcal{F}^2$$

$$H^1(C_2;I(E)) = \{1\}. \quad \blacksquare$$

With this observation we can prove

(3.3) Theorem: The group $\mathrm{Gen}(E/F)$ is an elementary abelian 2-group. It is trivial if no more than one prime ramifies. If $N > 1$ is the total number of primes, both finite and infinite, in F that ramify then $\mathrm{Gen}(E/F)$ has order 2^{N-1}. Furthermore the homomorphism $\mathrm{Gen}(E/F) \to H^2(C_2;E^*)$ given by $\langle y,A\rangle \to c\ell(y)$ is a monomorphism.

Proof:

Let us begin with the second part. So assume $y = z\bar{z}$ for some $z \in E^*$. Then $(zA)(\overline{zA}) = O_E$. But $H^1(C_2;I(E)) = \{1\}$ so for some fractional O_E-ideal B, $zA = B\bar{B}^{-1}$ or $A = z^{-1}B\bar{B}^{-1}$. Thus $\langle y,A\rangle = \langle 1,O_E\rangle$.

Now using (I.3.1) together with the realization of Hilbert symbols, subject to reciprocity, we may characterize the image of

$$1 \to \mathrm{Gen}(E/F) \to H^2(C_2;E^*)$$

as the subgroup of $c\ell(y)$ for which $h_P(c\ell(y)) = 1$ for all P inert. By reciprocity then

$$\prod_{\text{ramified}} h_P(c\ell(y)) = 1.$$

Hence the theorem. $\blacksquare$

Obviously there is a natural homomorphism

$$i_2\colon H^2(C_2;O_E^*) \to \mathrm{Gen}(E/F)$$

which to $u \in O_P^*$ assigns $\langle u,O_E\rangle$. If $v\bar{v} = u$ for some $v \in O_E^*$ then $(u,O_E) = (v\bar{v}, v^{-1}O_E\bar{O}_E\bar{v}^{-1})$, so i_2 is well defined.

To motivate our next step, let us prove i_2 can have a non-trivial kernel. Let $F = Q$ and $E = Q(\sqrt{34})$. Then -1 is the norm of $\frac{1}{3}(5+\sqrt{34})$, however the fundamental unit is $35 + 6\sqrt{34}$ and it has norm $+1$. Thus in this example $H^2(C_2;O_E^*)$ is cyclic of order 2 generated by $c\ell(-1)$ and $i_2(c\ell(-1)) = \langle -1,O_E\rangle = \langle 1,O_E\rangle$. Thus a global unit may be a norm from E^* without being a norm from O_E^*.

To explain this kernel we turn to the ideal class group $C(E)$ of the field E. Certainly this is still a C_2-module. An element of $H^2(C_2;C(E))$ arises from a O_E-ideal which is equivalent to its conjugate ideal, that is,

$$A\bar{A}^{-1} = z\, O_E$$

for some $z \in E^*$. But observe that $\bar{A}A^{-1} = \bar{z}\, O_E = (A\bar{A}^{-1})^{-1} = z^{-1}\, O_E$ and therefore $z\bar{z} \in O_F^*$. If $|A| \in C(E)$ denotes the ideal class of A then a homomorphism

$$d_2\colon H^2(C_2;C(E)) \to H^2(C_2;O_E^*)$$

is defined by $d_2(c\ell(|A|)) = c\ell(z\bar{z})$. If we can also write $A\bar{A}^{-1} = z_1\, O_E$ then $z_1 = zv$ for some $v \in O_E^*$ and $z_1\bar{z}_1 = z\bar{z}v\bar{v}$ and hence $c\ell(z_1\bar{z}_1) = c\ell(z\bar{z})$. If A were replaced by an equivalent ideal xA then

$$xA(\bar{x}^{-1}\bar{A}^{-1}) = x\bar{x}^{-1}\, z\, O_E$$

and $(x\bar{x}^{-1}z)\,\overline{(x\bar{x}^{-1}z)} = z\bar{z}$. Finally if $c\ell(|A|) = 1$ then $A = x\, B\bar{B}$ for some $x \in E^*$

and fractional O_E-ideal B. But $(x\,B\bar{B})(\bar{x}^{-1}\bar{B}^{-1}B^{-1}) = x\bar{x}^{-1}O_E$ and $(x\bar{x}^{-1})(\bar{x}x^{-1}) = 1$. Thus we do in fact have a well defined homomorphism.

(3.4) Lemma: The sequence

$$H^2(C_2;C(E)) \xrightarrow{d_2} H^2(C_2;O_E^*) \xrightarrow{i_2} \mathrm{Gen}(E/F)$$

is exact at the middle term.

Proof:

From the definition of d_2 and (I.3.3) it is immediate that composition is trivial. If for $u \in O_F^*$ we had $\langle u,O_E\rangle = \langle 1,O_E\rangle$ then we could write $u = z\bar{z}$, $O_E = z^{-1}\,B\bar{B}^{-1}$ and so $d_2(c\ell(|B|)) = c\ell(u)$. ■

It is quite easy to see that i_2 can have a cokernel. Let m be any square free integer. Let $F = Q$ and $E = Q(\sqrt{m})$. Then $H^2(C_2;O_E^*)$ is at most a cyclic group of order 2. But look at the order of $\mathrm{Gen}(E/F)$ from (I.3.3). Every prime dividing m ramifies. If $m \equiv 3 \pmod 4$ then 2 also ramifies and for $m < 0$ the real infinite prime in Q ramifies.

To understand the cokernel of i_2 we introduce $j_2\colon \mathrm{Gen}(E/F) \to H^1(C_2;C(E))$. For a pair (y,A) with $yA\bar{A} = O_E$ we see immediately that there is a $c\ell(|A|)$ in $H^1(C_2;C(E))$. If $(y,A) = (z\bar{z}, z^{-1}\,B\bar{B}^{-1})$ then $zA = B\bar{B}^{-1}$ and $c\ell(|A|) = 1$. Therefore we do receive a homomorphism $j_2\colon \mathrm{Gen}(E/F) \to H^1(C_2;C(E))$. As we expect

(3.5) Lemma: The sequence

$$H^2(C_2;O_E^*) \xrightarrow{i_2} \mathrm{Gen}(E/F) \xrightarrow{j_2} H^1(C_2;C(E))$$

is exact at the middle term.

Proof:

Composition is obviously trivial. Thus for $\langle y,A\rangle$ assume $c\ell(|A|) = 1 \in H^1(C_2;C(E))$. Then $zA = B\bar{B}^{-1}$ for some $z \in E^*$, B a fractional O_E-ideal. Now

$$y(zA)(\overline{zA}) = y\,O_E = z\bar{z}\,O_E$$

and therefore $y = z\bar{z}u$ for some $u \in O_F^*$ while $A = z^{-1}\,B\bar{B}^{-1}$. Thus $\langle y,A\rangle = \langle u,O_E\rangle$ as required. ■

The process may be carried one step further, which we shall need in the next section, by defining $d_1\colon H^1(C_2;C(E)) \to H^1(C_2;O_E^*)$ by analogy with d_2. Suppose $A\bar{A} = z\,O_E$, then $\bar{z}\,O_E = z\,O_E$ and hence $z\bar{z}^{-1} \in O_E^*$. We put $d_1(c\ell(|A|)) = c\ell(z\bar{z}^{-1})$ in $H^1(C_2;O_E^*)$. This is readily seen to be well defined.

(3.6) Lemma: The sequence

$$\text{Gen}(E/F) \xrightarrow{j_2} H^1(C_2;C(E)) \xrightarrow{d_1} H^1(C_2;O_E^*)$$

is exact at the middle term.

Proof:

Composition is trivial for if $A\bar{A} = y^{-1}\,O_E$, $y \in F^*$ we get $1 \in O_E^*$. Conversely assume $A\bar{A} = z\,O_E$ and $c\ell(z\bar{z}^{-1}) = 1$. We can write $z\bar{z}^{-1}v\bar{v}^{-1} = 1$ for some unit $v \in O_E^*$. But if $y^{-1} = zv$ then $\bar{y} = y \in F^*$ and $y\,A\bar{A} = O_E$ so that $j_2\langle y,A\rangle = c\ell(|A|)$. ■

So far there is an exact sequence

$$H^2(C_2;C(E)) \xrightarrow{d_2} H^2(C_2;O_E^*) \xrightarrow{i_2} \text{Gen}(E/F) \xrightarrow{j_2} H^1(C_2;C(E)) \xrightarrow{d_1} H^1(C_2;O_E^*)$$

which will close up to form an exact hexagon in the next section.

4. The group Iso(E/F)

By analogy with Gen(E/F) we shall define the group Iso(E/F). This time begin with pairs (u,K) wherein $u \in O_E^*$ is a unit with $u\bar{u}=1$ and $K \subset E$ is a fractional O_E-ideal for which $K=\bar{K}$. The pairs form a group with $(u,K)(u_1,K_1)=(uu_1,KK_1)$, identity $(1,O_E)$ and inverses (u^{-1},K^{-1}).

An equivalence relation between such pairs is introduced by $(u,K) \sim (u_1,K_1)$ if and only if there is $x \in E^*$ and a fractional ideal $A \subset E$ for which

$$x A\bar{A} K = K_1$$
$$x\bar{x}^{-1} u = u_1 .$$

Let us note $x A\bar{A} K = K_1$ implies $\bar{x} A\bar{A} = x A\bar{A}$ since $K=\bar{K}$, $K_1=\bar{K}_1$ and therefore $x\bar{x}^{-1}$ is a unit. This is readily seen to be an equivalence relation which preserves products and thus Iso(E/F) is the resulting group of equivalence classes. An element is denoted by $|u,K|$.

The group Iso(E/F) occurs in connection with the Witt group $H_u(K)$ for u-Hermitian forms with values in K. This Witt group is determined by $|u,K| \in \mathrm{Iso}(E/F)$. It was shown by Stoltzfus in [St, sec. 4] that if $\mathcal{D}_{E/Q}^{-1}$ is the inverse different of E/Q then the Witt groups $H_{\pm 1}(\mathcal{D}_{E/Q}^{-1})$ are significantly related to knot concordance as well as to cobordism of diffeomorphisms of closed manifolds.

We turn now to computing Iso(E/F). Given (u,K), when is $|u,K| = |1,O_E|$? We must produce a $z \in E^*$ and a fractional O_E-ideal A with

$$z\bar{z}^{-1} = u$$
$$z A\bar{A} = K.$$

We claim it is enough to produce $z\bar{z}^{-1} = u$ with $\nu_{\mathcal{P}}(z) \equiv \nu_{\mathcal{P}}(K) \pmod 2$ for all finite $\mathcal{P}=\bar{\mathcal{P}}$. If $\mathcal{P} \neq \bar{\mathcal{P}}$ there is no problem for since $z\bar{z}^{-1} = u \in O_E^*$ we will have

$\mathrm{ord}_{\mathscr{P}}\, z = \mathrm{ord}_{\overline{\mathscr{P}}}\, \overline{z} = \mathrm{ord}_{\overline{\mathscr{P}}}\, z$ and because $K = \overline{K}$, $\mathrm{ord}_{\mathscr{P}}\, K = \mathrm{ord}_{\overline{\mathscr{P}}}\, \overline{K} = \mathrm{ord}_{\overline{\mathscr{P}}}\, K$. Thus we could manufacture then an A with $z A\overline{A} = K$.

Assume now that at least one prime, finite or infinite, ramifies in E/F. By Hilbert Theorem 90 we can start with a $z\overline{z}^{-1} = u$. Using that ramified prime we can find a $t \in F^*$ so that for every inert P

$$(t,\sigma)_P = (-1)^{\nu_P(t)} = (-1)^{-\nu_{\mathscr{P}}(z) + \nu_{\mathscr{P}}(K)}.$$

Replace z by zt. Now we can assume $z\overline{z}^{-1} = u$ and $\nu_{\mathscr{P}}(z) \equiv \nu_{\mathscr{P}}(K) \pmod 2$ at all $\mathscr{P} = \overline{\mathscr{P}}$ over inert. We have just shown

(4.1) Lemma: <u>If at least one infinite prime ramifies, but no finite prime ramifies, then</u> $\mathrm{Iso}(E/F) = \{1\}$. ■

But what if finite primes ramify? Let $\mathscr{P} = \overline{\mathscr{P}}$ be such, then since $u\overline{u} = 1$ we have $c\ell(u) \in H^1(C_2; O_E(\mathscr{P})^*)$ and $c\ell(u) = c\ell(\pi\overline{\pi}^{-1})^{\nu_{\mathscr{P}}(z)}$ for any z with $z\overline{z}^{-1} = u$. Therefore to finish showing $|u,K| = |1,O_E|$ we would find that the necessary and sufficient condition is

$$c\ell(u) = c\ell(\pi\overline{\pi}^{-1})^{\nu_{\mathscr{P}}(K)} \in H^1(C_2; O_E(\mathscr{P})^*)$$

for all $\mathscr{P} = \overline{\mathscr{P}}$ over ramified.

If ramified finite primes are present let $\mathscr{R} \subset I(E)$ be the subgroup generated by all $\mathscr{P} = \overline{\mathscr{P}}$ over ramified. We define a homomorphism $\mathrm{Iso}(E/F) \to \mathscr{R}/\mathscr{R}^2$ by associating to $|u,K|$ the product

$$\prod_{\text{ramified}} \mathscr{P}^{\nu_{\mathscr{P}}(K) - \nu_{\mathscr{P}}(z)}$$

mod $\mathcal{R}^2$ where $z\bar{z}^{-1} = u$. This is well defined for if

$$x A\bar{A} K = K_1$$
$$x\bar{x}^{-1} u = u_1$$

we would choose $z_1 = xz$, but

$$\nu_{\mathcal{P}}(x) + \nu_{\mathcal{P}}(K) \equiv \nu_{\mathcal{P}}(K_1) \pmod 2$$

so $\nu_{\mathcal{P}}(K) - \nu_{\mathcal{P}}(z_1) = \nu_{\mathcal{P}}(K_1) - \nu_{\mathcal{P}}(x) - \nu_{\mathcal{P}}(z) \equiv \nu_{\mathcal{P}}(K) - \nu_{\mathcal{P}}(z) \pmod 2$.

Our argument shows this is a monomorphism. Look at $|1,\mathcal{P}|$ for $\mathcal{P} = \bar{\mathcal{P}}$ over ramified to see it is also an epimorphism.

(4.2) __Lemma__: __If finite ramified primes occur in__ E/F __then__ $\mathrm{Iso}(E/F) \simeq \mathcal{R}/\mathcal{R}^2$. ∎

One case is now left. We say that E/F is __unramified__ if and only if no prime of F, finite or infinite ramifies. We begin by returning to the group of pairs (u,K). From this group we define a homomorphism into Z^* by choosing a $z \in E^*$ with $z\bar{z}^{-1} = u$ and sending

$$(u,K) \to (-1)^{\Sigma \nu_{\mathcal{P}}(K) + \nu_{\mathcal{P}}(z)}$$

where the sum extends over all $\mathcal{P} = \bar{\mathcal{P}}$ over inert. To see this is well defined suppose $z_1\bar{z}_1^{-1} = u$ also. Then $zt = z_1$ for some $t \in F^*$ and $\nu_{\mathcal{P}}(K) + \nu_{\mathcal{P}}(z_1) = \nu_{\mathcal{P}}(K) + \nu_{\mathcal{P}}(z) + \nu_P(t)$. By reciprocity

$$\Pi\ (t,\sigma)_P = (-1)^{\Sigma \nu_P(t)} = 1$$

where product (and sum) runs over all P inert. Thus the homomorphism is well defined.

There is now no problem in showing that there is induced

$$\mathrm{Iso}(E/F) \to Z^* .$$

If $x A\bar{A} K = K_1$, $x\bar{x}^{-1} u = u_1$ then $\mathrm{ord}_{\mathscr{P}} x + \mathrm{ord}_{\mathscr{P}} K \equiv \mathrm{ord}_{\mathscr{P}} K_1 \pmod 2$ for all $\mathscr{P} = \bar{\mathscr{P}}$ over inert.

(4.3) <u>Lemma</u>: <u>If</u> E/F <u>is unramified then</u> $\mathrm{Iso}(E/F) \simeq Z^*$.

<u>Proof</u>:

If $(-1)^{\Sigma \nu_{\mathscr{P}}(K) + \nu_{\mathscr{P}}(z)} = 1$ then we can find $t \in F^*$ so that for every inert P

$$(t,\sigma)_P = (-1)^{\nu_P(t)} = (-1)^{\nu_{\mathscr{P}}(K) + \nu_{\mathscr{P}}(z)} .$$

It is the assumption that

$$\sum \nu_{\mathscr{P}}(K) + \nu_{\mathscr{P}}(z) \equiv 0 \pmod 2$$

which allows us to so assign Hilbert symbols and still satisfy reciprocity. Replacing z by zt we have $z\bar{z}^{-1} = u$ and $\mathrm{ord}_{\mathscr{P}} z \equiv \mathrm{ord}_{\mathscr{P}} K \pmod 2$ for all $\bar{\mathscr{P}} = \bar{\mathscr{P}}$ over inert. The rest of the argument is as before. In this case the generator of $\mathrm{Iso}(E/F)$ is $|1,\mathscr{P}|$ where $\mathscr{P} = \bar{\mathscr{P}}$ is any prime over an inert.

We can now complete the exact hexagon. Define

$$i_1 : H^1(C_2; O_E^*) \to \mathrm{Iso}(E/F)$$

by

$$i_1(c\ell(u)) = |u, O_E|$$

and $j_1 : \mathrm{Iso}(E/F) \to H^2(C_2; C(E))$ by

$$j_1(|u,K|) = c\ell(|K|).$$

(4.4) Theorem: The hexagon

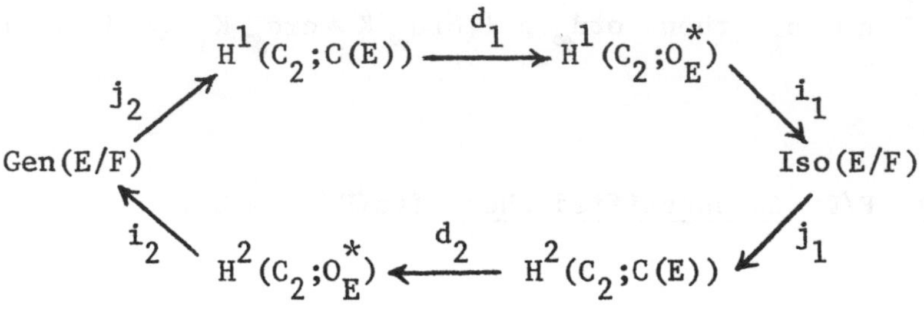

is exact.

Proof:

In view of our discussion in I.3 we shall only take up $\text{im}(j_1) = \ker(d_2)$. Since $\overline{K} = K$ we have $K\overline{K}^{-1} = 1 \cdot O_E$ and thus the composition $d_2 \circ j_1$ is trivial.

Suppose now $d_2(c\ell(|A|)) = 1 \in H^2(C_2;O_E^*)$. Then $A\overline{A}^{-1} = z\,O_E$ and for some $v \in O_E^*$ we have $z\overline{z}v\overline{v} = 1$. So write $w = zv$ and then $A\overline{A}^{-1} = w\,O_E$ while $w\overline{w} = 1$. Use Hilbert Theorem 90 to write $x\overline{x}^{-1} = w$ and note $\overline{(x^{-1}A)} = x^{-1}A$. Therefore $j_1(|1,x^{-1}A|) = c\ell(|A|)$ in $H^2(C_2;C(E)^*)$. ■

The exactness itself is nothing but a consequence of definitions. What is of concern are the groups Gen(E/F) and Iso(E/F). In determining these it will be noted that we used Hilbert symbols and the cyclic norm theorem. It is in those computations that we are able to utilize algebraic number theory's powerful technique of localization and completion. By contrast such techniques are not applicable to the determination of $H^*(C_2;C(E))$ and $H^*(C_2;O_E^*)$. We especially caution our reader that to rephrase in terms of idele class groups, a step we have intentionally avoided, will not eliminate any problem.

In the next sections we will present a number of elementary examples to illustrate our point.

5. The unramified case

We open this section with a general comment about the relation of the ideal class groups $C(F)$ and $C(E)$. There is a natural homomorphism $C(F) \to C(E)$ induced by sending each fractional O_F-ideal $\mathfrak{a} \subset F$ into the fractional O_E-ideal $\mathfrak{a}\cdot O_E$. We denote the kernel of this homomorphism by $K(E/F) \subset C(F)$.

We shall also use the extension of the norm homomorphism to ideals. This is the homomorphism $N: I(E) \to I(F)$ which on generators has the effect:

$$N(\mathscr{P}) = P \quad \text{if} \quad \mathscr{P} \neq \bar{\mathscr{P}} \quad \text{is over split}$$

$$N(\mathscr{P}) = P^2 \quad \text{if} \quad \mathscr{P} = \bar{\mathscr{P}} \quad \text{is over inert}$$

$$N(\mathscr{P}) = P \quad \text{if} \quad \mathscr{P} = \bar{\mathscr{P}} \quad \text{over ramified.}$$

We shall make particular use of the relation $N(\mathfrak{a}\cdot O_E) = \mathfrak{a}^2$ for any fractional O_F-ideal.

(5.1) Lemma: <u>There is an isomorphic embedding of $K(E/F)$ into $H^1(C_2;O_E^*)$. If no finite prime ramifies then $K(E/F) \simeq H^1(C_2;O_E^*)$. If finite primes ramify then</u>

$$1 \to K(E/F) \to H^1(C_2;O_E^*) \xrightarrow{i_1} \mathrm{Iso}(E/F)$$

<u>is exact.</u>

<u>Proof</u>:

Suppose $\mathfrak{a} \subset F$ is a fractional O_F ideal for which $\mathfrak{a}\cdot O_E = z\,O_E$ for some $z \in E^*$. Then $\bar{z}\,O_E = z\,O_E$ so $z\bar{z}^{-1} \in O_E^*$. If we wrote $z_1 \cdot O_E = \mathfrak{a}\cdot O_E$ then $z_1 = zv$ for $v \in O_E^*$ and $c\ell(z_1\bar{z}_1^{-1}) = c\ell(z\bar{z}^{-1}) \in H^1(C_2;O_E^*)$. If $\mathfrak{a}$ were replaced by an equivalent ideal $y\mathfrak{a}$, $y \in F^*$, then $y\mathfrak{a}\cdot O_E = yz\,O_E$ but $(yz)(\overline{yz})^{-1} = z\bar{z}^{-1}$. Thus we do obtain a well defined homomorphism $K(E/F) \to H^1(C_2;O_E^*)$.

To see that this is an embedding assume $\mathfrak{a}\cdot O_E = z\,O_E$ and $c\ell(z\bar{z}^{-1}) = 1$. Then there is a unit $v \in O_E^*$ with $z\bar{z}^{-1}v\bar{v}^{-1} = 1$. Let $y = zy$ so $y = \bar{y} \in F^*$

and $\mathfrak{a}\cdot O_E = y\cdot O_E$. Applying the norm homomorphism, $\mathfrak{a}^2 = y^2 O_F$. Since $I(F)$ is a free abelian group, $\mathfrak{a} = y O_F$.

Next we must find the image. Suppose $u \in O_E^*$ is a unit with $u\bar{u} = 1$. Write $z\bar{z}^{-1} = u$. Since u is a unit we must have

$$\mathrm{ord}_{\mathscr{P}}\, z - \mathrm{ord}_{\mathscr{P}}\, \bar{z} = \mathrm{ord}_{\mathscr{P}}\, z - \mathrm{ord}_{\bar{\mathscr{P}}}\, z = 0$$

for every $\mathscr{P} \subset O_E$. Surely we can find a fractional O_F-ideal $\mathfrak{a} \subset F$ with

$$\mathrm{ord}_P\, \mathfrak{a} = \mathrm{ord}_{\mathscr{P}}\, z, \quad P \text{ inert}$$

$$\mathrm{ord}_P\, \mathfrak{a} = \mathrm{ord}_{\mathscr{P}}\, z = \mathrm{ord}_{\bar{\mathscr{P}}}\, z, \quad P \text{ split.}$$

If no finite prime ramifies then $\mathfrak{a}\cdot O_E = z\cdot O_E$ and $K(E/F) \simeq H^1(C_2;O_E^*)$.

But if finite primes ramify we would need to know $\mathrm{ord}_{\mathscr{P}}\, z \equiv 0 \pmod 2$ for all $\mathscr{P} = \bar{\mathscr{P}}$ over ramified so that we could have $2\, \mathrm{ord}_P\, \mathfrak{a} = \mathrm{ord}_{\mathscr{P}}\, z$ for all P ramified. Hence the lemma. ■

The group $K(E/F)$ is always an elementary abelian 2-group. It can be trivial in which case

$$i_1\colon H^1(C_2;O_E^*) \to \mathrm{Iso}(E/F)$$

will be a monomorphism. However,

(5.2) <u>Lemma</u>: <u>If</u> E/F <u>is unramified then</u> $K(E/F)$ <u>is non-trivial</u>. <u>Furthermore</u> $i_1\colon H^1(C_2;O_E^*) \to \mathrm{Iso}(E/F) \simeq Z^*$ <u>is non-trivial if and only if there is an inert prime</u> $P \subset O_F$ <u>for which</u> $\mathscr{P} = P\cdot O_E$ <u>is principal</u>.

<u>Proof</u>:

In this situation by (I.3.3), $\mathrm{Gen}(E/F)$ is trivial while by (I.4.3) $\mathrm{Iso}(E/F) \cong Z^*$. If $K(E/F) = \{1\}$, then by (I.5.1) so is $H^1(C_2;O_E^*)$. From (I.4.4) it would follow $H^1(C_2;C(E))$ is also trivial. But $C(E)$ is a finite group so $H^1(C_2;C(E)) \simeq H^2(C_2;C(E)) = \{1\}$. This contradicts (I.4.4) since $\mathrm{Iso}(E/F) \simeq Z^*$.

The next part is a bit more delicate. We should think in terms of the Artin map [L, p. 197]. Initially this is the homomorphism which sends $I(F)$ onto Z^* by assigning -1 to each inert prime and 1 to each split prime. That it takes value 1 on principal O_F-ideals is a consequence of the Hilbert reciprocity. Thus there is induced

$$\omega: C(F) \to Z^*.$$

Now remember $\mathrm{Iso}(E/F) \simeq Z^*$ with generator $|1,\mathcal{P}|$ where $\mathcal{P} = \bar{\mathcal{P}}$ is over a ramified. Thus the Artin map ω is the composition of

$$C(F) \to \mathrm{Iso}(E/F) \simeq Z^*$$

where $\mathfrak{a} \to |1, \mathfrak{a}\cdot O_E|$. In particular, then

$$K(E/F) \simeq H^1(C_2;O_E^*) \xrightarrow{i_1} \mathrm{Iso}(E/F) \simeq Z^*$$

is the restriction $\omega|K(E/F) \to Z^*$. The second part of the lemma will now follow if we recall that every ideal class contains a prime ideal. ∎

More generally the existence of relative quadratic unramified extensions of a field F may be predicted from class field theory. If H/F is the maximal abelian unramified extension of F, the Hilbert Class Field of F [L, p. 224], then Artin's reciprocity map provides an explicit isomorphism

$$\omega: C(F) \simeq G(H/F)$$

between the class group of F and the Galois group of H/F. Thus to every subgroup of index 2 in $C(F)$ there will correspond an unramified quadradic extension of F.

For example let $F = Q(\sqrt{-39})$ so that $C(F) \simeq C_4$, [C, p. 264]. There will be a unique unramified quadratic extension (constructed below), E/F. Now $K(E/F)$ is a non-trivial 2-group so we have it identified here and $\omega|K(E(F) \to Z^*$ is surely trivial. Then by (I.5.2) and (I.4.4) we find $d_1: H^1(C_2;C(E)) \simeq H^1(C_2;O_E^*) \simeq C_2$ and $Iso(E/F) \simeq H^2(C_2;C(E)) \simeq C_2$ while $H^2(C_2;O_E^*) = \{1\}$ in this example.

Here is the classical construction of unramified quadratic extensions. Let m_1, m_2 be relatively prime square free integers. Assume $m_1 \equiv 1 \pmod 4$ and that m_1, m_2 are not both negative. Let $F = Q(\sqrt{m_1 m_2})$ and then $E = F(\sqrt{m_1}) = F(\sqrt{m_2})$ is an unramified quadratic extension of F. The explanation is this. The field E is the compositum of $Q(\sqrt{m_1})$ and $Q(\sqrt{m_2})$. But $dis(Q(\sqrt{m_1}/Q)$ is relatively prime to $dis(Q(\sqrt{m_2}/Q)$. Hence, [L, p. 68], the discriminant of E/Q is $(dis\ Q(\sqrt{m_1})/Q)^2 (dis\ Q(\sqrt{m_2})/Q)^2 = dis(Q(\sqrt{m_1 m_2}/Q)^2$. This insures that in E/F no finite prime could possibly ramify. We assumed m_1 and m_2 are not both negative which guarantees that no infinite prime in F ramifies.

Every textbook mentions $m_1 = 5$, $m_2 = -1$. In this case $Q(\sqrt{-5}) = F$ has class group C_2 so that in fact E/F is the Hilbert class field of F, $K(E/F) = C(F)$ and

$$i_1: H^1(C_2;O_E^*) \simeq Iso(E/F).$$

From (I.4.4) then in this example

$$H^1(C_2;C(E)) \simeq H^2(C_2;C(E)) \simeq H^2(C_2;O_E^*) = \{1\}.$$

This should be contrasted with $m_1 = -3$, $m_2 = 13$ which produces the unramified quadratic extension of $Q(\sqrt{-39})$ mentioned earlier. This is not the Hilbert class field of $Q(\sqrt{-39})$.

There is always $c\ell(-1) \in H^1(C_2;O_E^*)$. If $m_1 m_2 < 0$ we can say when $1 = c\ell(-1)$. The point is that with this assumption O_F^*, the group of units for $Q(\sqrt{m_1 m_2})$, only consists of 1 and -1. Thus if $c\ell(-1) = 1$ there is a unit $v \in O_E^*$ with

$\bar{v} = -v$ and $v\bar{v} = -v^2 \in O_F^*$. If $-v^2 = -1$ then $v = \pm 1$ which is a contradiction. Thus $-v^2 = 1$ or $v = \sqrt{-1}$. That is, E contains $Q(\sqrt{-1})$. Yet E/Q is Galois with group the Klein viergruppe so its quadratic subfields are $Q(\sqrt{m_1})$, $Q(\sqrt{m_2})$ and $Q(\sqrt{m_1 m_2})$. Therefore if $m_1 m_2 < 0$, $c\ell(-1) = 1 \in H^1(C_2;O_E^*)$ if and only if $m_2 = -1$.

This was used in [St, sec. 4] as follows. Consider $m_1 = -3$, $m_2 = 5$. Then $C(F) \simeq C_2$ for $F = Q(\sqrt{-15})$ and with $E = F(\sqrt{-3}) = F(\sqrt{5})$ we find $+1 \neq c\ell(-1) \in H^1(C_2;O_E^*)$ and $i_1(c\ell(-1)) = |-1,O_E| \in \mathrm{Iso}(E/F) \simeq Z^*$ is the generator. The composition

$$H^1(C_2;O_E^*) \xrightarrow{i_1} \mathrm{Iso}(E/F) \simeq Z^*$$

is, in the unramified case, the homomorphism $Sc: H^1(C_2;O_E^*) \to Z^*$ of [St, p. 53]. We shall continue to use Sc.

6. Examples

The most elementary example which may be used to illustrate the groups and homomorphisms of (I.4.4) is $F = Q$ and $E = Q(\sqrt{m})$ where $m > 0$ is a square free integer. In this case the infinite prime is unramified but every finite rational prime dividing m ramifies. If $m \equiv -1 \pmod 4$ then in addition 2 will ramify.

Thus we are in the situation of (I.4.2) and $\mathrm{Iso}(E/F) \simeq \mathcal{R}/\mathcal{R}^2$ is an elementary abelian 2-group whose order is 2^n where n is the number of rational ramified primes. Furthermore, by (I.3.3) Gen(E/F) is an elementary abelian group of order 2^{n-1}. Since Q has class number 1 we also find from (I.5.1) that $i_1: H^1(C_2;O_E^*) \to \mathrm{Iso}(E/F)$ is a monomorphism. Under $\mathrm{Iso}(E/F) \simeq \mathcal{R}/\mathcal{R}^2$ the image $i_1(c\ell(-1))$ is $\mathfrak{D}_{E/F} \pmod{\mathcal{R}^2}$. Actually this is always non-trivial in these examples (there is always a finite ramified prime of type I). Thus $c\ell(-1) \neq 1 \in H^1(C_2;O_E^*)$.

(6.1) Lemma: If $E = Q(\sqrt{m})$ where $m > 0$ square free integer then

1) <u>if -1 is the norm of a unit in</u> O_E^*

$$H^2(C_2;O_E^*) = \{1\}$$
$$H^1(C_2;O_E^*) \simeq C_2$$

2) <u>if -1 is not the norm of a unit in</u> O_E^*

$$H^2(C_2;O_E^*) \simeq C_2$$
$$H^1(C_2;O_E^*) \simeq C_2 \oplus C_2$$

3) $j_2\colon \mathrm{Gen}(E/F) \to H^1(C_2;C(E))$ <u>is an epimorphism</u>

4) $i_2\colon H^2(C_2;O_E^*) \to \mathrm{Gen}(E/F)$ <u>is trivial if and only if every odd rational prime</u> $p|m$ <u>is congruent to</u> 1 (mod 4).

<u>Proof:</u>

Part 3) follows from (I.4.4) since i_1 is a monomorphism. Since $O_F^* = \{-1,1\}$ the statements about $H^2(C_2;O_E^*)$ are clear. We use (I.2.2) to find when -1 is a norm from E^*. If 2 is the only ramified prime, $m = 2$, then $(-1,2)_2 = 1$ by reciprocity since -1 is a unit. If $p|m$ is odd then $(-1,m)_p = 1$ if and only if $-1 \in F_p$, the residue field, is a square (I.2.4); that is, if and only if $p \equiv 1 \pmod 4$. If $2|m$ and all odd primes dividing m are 1 (mod 4) then $(-1,m)_2 = 1$ by reciprocity. This establishes part 4).

Returning to 1) and 2) we have a short exact sequence of C_2-modules

$$1 \to O_F^* \to O_E^* \to O_E^*/O_F^* \to 1.$$

By the Dirichlet units theorem O_E^*/O_F^* is a multiplicatively written infinite cyclic group. Furthermore, for any unit $u \in O_E^*$, $u\bar{u} = \pm 1$. Hence $H^2(C_2;O_E^*/O_F^*) = \{1\}$ and $H^1(C_2;O_E^*/O_F^*) \simeq C_2$. Using the derived long exact sequence

$$1 \to H^1(C_2;O_F^*) \to H^1(C_2;O_E^*) \to H^1(C_2;O_E^*/O_F^*)$$
$$\to H^2(C_2;O_F^*) \to H^2(C_2;O_E^*) \to 1$$

the assertions about $H^1(C_2;O_E^*)$ now follow. ■

As we noted earlier there are cases, such as $Q(\sqrt{34})$ where in -1 is a norm from E^* but not from O_E^*. However for $E = Q(\sqrt{p})$, p a prime congruent to 1 (mod 4), -1 is always the norm of a fundamental unit in O_E^*.

The next class of examples is more general. Let F be a completely real algebraic number field; that is, every infinite prime in F is real. We shall assume also that $\sigma \in F^*$ is totally negative; that is σ is negative for every ordering of F. When we form $E = F(\sqrt{\sigma})$ every real infinite prime in F ramifies.

Let Λ be the cyclic group of all roots of unity in O_E^*. Then $\Lambda \cap O_F^* = \{-1,1\}$. If $\lambda \in \Lambda$ then $\lambda\bar{\lambda} = 1$ for surely $\lambda\bar{\lambda}$ is still a root of unity and -1 cannot be a norm. Thus the involution on Λ is $\bar{\lambda} = \lambda^{-1}$. Obviously $H^2(C_2;\Lambda) \simeq C_2$ generated by $c\ell(-1)$. About $H^1(C_2;\Lambda)$, it is also C_2 and generated by $c\ell(\gamma)$ where $\gamma \notin \Lambda^2$.

Look now at the short exact sequence of C_2-modules

$$1 \to \Lambda \to O_E^* \to O_E^*/\Lambda \to 1.$$

The quotient is a multiplicatively written finitely generated free abelian group which contains $O_F^*/\{-1,1\}$ as a subgroup. By the Dirichlet units theorem [L, p. 105]

$$\text{rank } O_E^*/\Lambda = \text{rank } O_F^*/\{-1,1\} = r-1$$

where r is the number of infinite primes in F. Since the induced involution on O_E^*/Λ is the identity on $O_F^*/\{-1,1\}$ it follows that it must be the identity on O_E^*/Λ itself. Therefore

$$H^2(C_2;O_E^*/\Lambda) \simeq (C_2)^{r-1}$$

$$H^1(C_2;O_E^*/\Lambda) = \{1\}.$$

Furthermore, $H^2(C_2;\Lambda) \to H^2(C_2;O_E^*)$ is a monomorphism since -1 cannot be a norm.

(6.2) <u>Lemma</u>: <u>If E is the adjunction of the square root of a totally negative element to a completely real field then either</u>

<u>Case</u> A)

$$H^1(C_2;\Lambda) \simeq H^1(C_2;O_E^*) \simeq C_2$$

$$H^2(C_2;O_E^*) \simeq (C_2)^r$$

or

<u>Case</u> B)

$$H^1(C_2;O_E^*) \simeq \{1\}$$

$$H^2(C_2;O_E^*) \simeq (C_2)^{r-1}.$$

<u>Proof</u>:

This follows immediately from the exact sequence of C_2-cohomology groups associated to

$$1 \to \Lambda \to O_E^* \to O_E^*/\Lambda \to 1. \quad ■$$

If $O_F^{**} \subset O_F^*$ is the subgroup of squares of units then obviously there is an epimorphism

$$O_F^*/O_F^{**} \to H^2(C_2;O_E^*).$$

Now we may say

(6.3) Lemma: In case A of (I.6.2)

$$O_F^*/O_F^{**} \simeq H^2(C_2;O_E^*)$$

and every unit in O_E^* is the product of a unit in O_F^* with a root of unity. In case B of (I.6.2) the kernel of

$$O_F^*/O_F^{**} \to H^2(C_2;O_F^*) \to 1$$

is cyclic of order 2.

Proof:

For any unit $u \in O_E^*$ we can write $\bar{u} = u\lambda$ for some $\lambda \in \Lambda$. Since $u^{-1}\bar{u} = \lambda$ it follows that $c\ell(\lambda) \in H^1(C_2;\Lambda)$ lies in the kernel of $H^1(C_2;\Lambda) \to H^1(C_2;O_E^*)$. Thus in case A we can write $\lambda = \mu^2$ for some $\mu \in \Lambda$. Let $v = u\mu$, then $\overline{u\mu} = \bar{v} = \bar{u}\bar{\mu}^{-1} = u\lambda\mu^{-1} = u\mu = v$ so $v \in O_F^*$ and $v\mu^{-1} = u$. Also $u\bar{u} = v^2$ so that any unit which is a norm from O_E^* is already a square in O_F^*.

Turn now to case B. There is a $\gamma \notin \Lambda^2$ for which $c\ell(\gamma) \in H^1(C_2;\Lambda)$ is in the kernel of $H^1(C_2;\Lambda) \to H^1(C_2;O_E^*)$. Thus for some $u \in O_E^*$, $u^{-1}\bar{u} = \gamma$ or $\bar{u} = u\gamma$. Then $u\bar{u} = u^2\gamma \in O_F^*$. But if $u^2\gamma = v^2$ for some $v \in O_F^*$ we would find $\gamma = (vu^{-1})^2$. But then $vu^{-1} \in \Lambda$ and $\gamma \in \Lambda^2$, a contradiction. Thus $u^2\gamma$ is a norm from O_E^* but is not a square in O_F^*. ■

Refer back to (I.5.1) to see

(6.4) Lemma: In case B of (I.6.2) K(E/F) is trivial.

The most common example of case B arises in the cyclotomic number field $Q(\xi_n)$ where n is an odd composite number. The fixed field under complex conjugation, $Q(\xi_n + \xi_n^{-1})$, is completely real and $\varphi(n)/2$ is the number of its orderings. In this case Λ has order $4k+2$ so $c\ell(-1)$ generates $H^1(C_2;\Lambda)$.

However $\xi_n - \xi_n^{-1}$ is a unit so that $c\ell(-1) = 1 \in H^1(C_2;O_E^*)$. We refer to [We, p. 255] for more comments about cyclotomic number fields.

(6.5) Lemma: If E is the adjunction of the square root of a totally negative element to a completely real field and if at least one prime of type I ramifies then

1) $H^1(C_2;O_E^*) \simeq C_2$ generated by $c\ell(-1)$

2) $H^1(C_2;O_E^*) \to \mathrm{Iso}(E/F) \simeq \mathcal{R}/\mathcal{R}^2$ is a monomorphism with

$c\ell(-1) \to \mathcal{D}_{E/F} \pmod{\mathcal{R}^2}$

3) $K(E/F)$ is trivial.

Proof:

This is a consequence of the definition of a finite ramified prime of type I (I.1.4,1.7) together with case A of (I.6.2) and (I.5.1). ■

For $Q(\sqrt{m})/Q$ with $m < 0$ square free integer we are always in case A of (I.6.2). Only $\sqrt{-1}$ is not immediately covered by (I.6.5). In that case $c\ell(i)$ generates $H^1(C_2;O_E^*) \simeq C_2$.

On the subject of cyclotomic number fields (I.6.5) applies to complex conjugation on $Q(\xi_{p^k})$, p^k a power of a single odd prime.

We might also mention a result of Kummer when $K(E/F)$ is trivial. In that situation we have

$$1 \to C(F) \to C(E) \to C(E)/C(F) \to 1.$$

If h, h_1 and h_2 are respectively the orders of $C(E)$, $C(E)/C(F)$ and $C(F)$ then there is $h = h_1 h_2$.

(6.6) <u>Kummer</u>: <u>If</u> $K(E/F)$ <u>is trivial then</u> $h_2 \equiv 0 \pmod 2$ <u>implies</u> $h_1 \equiv 0 \pmod 2$.

<u>Proof</u>:

Since $K(E/F)$ is trivial, at least one prime, finite or infinite, ramifies. For any fractional ideal $\mathfrak{a} \subset F$ over O_F we then claim $c\ell(|\mathfrak{a}\cdot O_E|) = 1 \in H^2(C_2;C(E))$. In fact $|1,\mathfrak{a}\cdot O_E| = |1,O_E|$ in $\mathrm{Iso}(E/F)$. If no finite prime ramifies then $\mathrm{Iso}(E/F)$ is trivial. If finite ramified primes do occur then for $\mathcal{P} = \overline{\mathcal{P}}$ over ramified $\mathrm{ord}_{\mathcal{P}}\, \mathfrak{a}\cdot O_E \equiv 0 \pmod 2$ and by (I.4.2) $|1,\mathfrak{a}\cdot O_E| = |1,O_E|$.

The short exact sequence of C_2-modules $1 \to C(F) \to C(E) \to C(E)/C(F) \to 1$ then produces

$$\cdots \to H^1(C_2;C(E)/C(F)) \to H^2(C_2;C(F)) \to H^2(C_2;C(E)) \to \cdots$$

and $H^1(C_2;C(E)/C(F)) \to H^2(C_2;C(F)) \simeq C(F)/C(F)^2$ is an epimorphism by our preceeding remarks. Hence Kummer's result. ■

To us the most puzzling phenomenon is the kernel of $H^2(C_2;O_E^*) \to \mathrm{Gen}(E/F)$. This represents units in O_F^* which are norms from E^* but not norms from O_E^*. We shall say the little we know about this phenomenon, and mention a few examples. Let $O_F^+ \subset O_F^{**}$ be the subgroup of totally positive elements; that is, the units which are positive with respect to every ordering of F. Clearly $O_F^{**} \subset O_F^+$. Then we have

(6.7) <u>Lemma</u>: <u>Let</u> E <u>be the adjunction of the square root of a totally negative element to a completely real field</u> F. <u>If only one finite prime ramifies, and it is of type</u> I, <u>then</u>

$$d_2\colon H^2(C_2;C(E)) \to H^2(C_2;O_E^*)$$

<u>is an isomorphism onto the image of</u> O_F^+/O_F^{**} <u>under the isomorphism</u>

$$O_F^*/O_F^{**} \simeq H^2(C_2;O_E^*).$$

Proof:

Surely we are in case A and since only one prime in O_F ramifies $H^1(C_2;O_E^*) \simeq \mathrm{Iso}(E/F) \simeq C_2$. We claim that any unit $u \in O_E^+$ is a norm from E^*. If P_o is the only finite ramified prime then $(u,\sigma)_{P_o} = 1$ by reciprocity. Hence $\langle u,O_E\rangle = 1 \in \mathrm{Gen}(E/F)$ so that $c\ell(u)$ lies in the image of d_2. ■

The obvious question then is how do we recognize the non-triviality of O_F^+/O_F^{**}.

(6.8) Lemma: If F is completely real then $O_F^+ = O_F^{**}$ if and only if O_F^* contains units with arbitrarily prescribed signs.

Proof:

Let r be the number of (real) infinite primes in F. Define a homomorphism

$$O_F^* \to (Z^*)^r$$

by assigning to u a sequence with a -1 for each ordering where u is negative and a $+1$ for each ordering where u is positive. There will result

$$1 \to O_F^+/O_F^{**} \to O_F^*/O_F^{**} \to (Z^*)^r.$$

By the units theorem O_F^*/O_F^{**} is also an elementary abelian 2 group of order 2^r. Thus $O_F^*/O_F^{**} \to (Z^*)^r$ is an epimorphism, if and only if it is a monomorphism, if and only if it is an isomorphism. ■

Surely there are many completely real fields that do not contain units with arbitrarily prescribed signs. For example $Q(\sqrt{3})$ does not. On the other hand $Q(\sqrt{5})$ does. The following is a helpful guide.

(6.9) Lemma: Let K be a completely real field and let σ be a totally positive non-square in K. If $F = K(\sqrt{\sigma})$, also completely real, has units with

<u>arbitrarily prescribed signs then so does</u> K <u>and</u> $H^2(C_2;O_K^*) = \{1\}$.

<u>Proof</u>:

This time every ordering of K extends to a conjugate pair of orderings of F. Pick an ordering of K and one of the orderings of F to which it extends. By hypothesis there is a $u \in O_F^*$ negative at this ordering and positive at every other ordering of F. But then $\bar{u}$ is positive in the order where u was negative and thus $u\bar{u}$ is negative at the originally chosen order of K and positive for every other ordering of K.

Thus we have found units in O_K^* with arbitrarily prescribed signs and which are also norms from O_F^*. Now from $O_K^*/O_K^{**} \simeq (Z^*)^r$ it will follow that $H^2(C_2;O_K^*) = \{1\}$. ■

This is used as follows. We select a positive square free integer m divisible by at least one prime congruent to 3 (mod 4). Let $F = Q(\sqrt{m})$. Since -1 cannot be a norm it follows from (I.6.8, 6.9) that O_F^+/O_F^{**} is non-trivial (actually cyclic of order 2). Next we seek out a prime p so that

1) $p \equiv 3 \pmod 4$
2) $(p,m) = 1$
3) p is inert in $Q(\sqrt{m})/Q$.

Next let $E = F(\sqrt{-p})$. We see that $p\,O_F$ is the only finite prime to ramify in E/F. To see this write $-p = 1 + 4k$ so that $\theta = \frac{-1+\sqrt{-p}}{2}$ belongs to O_E and the discriminant of this element is -p. We apply (I.6.7) to see that

$$d_2\colon H^2(C_2;C(E)) \to H^2(C_2;O_E^*)$$

is non-trivial (image cyclic of order 2). To be completely specific let $F = Q(\sqrt{3})$ and $E = F(\sqrt{-7})$.

Actually Kummer by explicit computations exhibited the non-triviality of d_2 for $Q(\xi_{29})/Q(\xi_{29}+\xi_{29}^{-1})$. To close the section we suggest some exercises using (I.4.4) to establish some well-known results.

(6.10) **Exercise:** If the class number of F is odd show that $\tilde{C}(E) \subset C(E)$, the subgroup of elements of order ≤ 2, is naturally isomorphic to $H^2(C_2;C(E))$.

(6.11) **Exercise:** If the class number of F is odd and exactly one prime ramifies in E/F show that the class number of E is also odd.

For $k \geq 2$ let $\xi_{2^k} = \exp(2\pi i/2^k)$. Then there are the cyclotomic number fields $Q(\xi_{2^k})$.

(6.12) **Exercise:** Show that the class number of $Q(\xi_{2^k})$ is odd.

II The Witt Ring H(E)

1. General definitions

Again E/F is a relative quadratic extension. We wish to describe the Witt ring of Hermitian innerproduct spaces over E. A <u>Hermitian innerproduct space</u> is a pair (V,h) consisting of a finite dimensional vector space over E together with a biadditive function

$$h: V \times V \to E$$

satisfying $\lambda h(v,w) = h(\lambda v,w) = h(v,\bar{\lambda} w)$ and $h(w,v) = \overline{h(v,w)}$. Furthermore it is assumed that

$$\mathrm{Ad}: V \to \mathrm{Hom}_E(V,E)$$

given by $w \to h(\cdot,w)$ is an isomorphism of E vector spaces. It must, however, be understood that scalar products on $\mathrm{Hom}_E(V,E)$ are defined by

$$(\lambda\varphi)(v) \equiv \varphi(\bar{\lambda} v)$$

for $\lambda \in E$, $v \in V$ and $\varphi \in \mathrm{Hom}_E(V,E)$.

The <u>orthogonal sum</u> of two Hermitian innerproducts spaces (V,h) and (V_1,h_1) is obtained by forming $V \oplus V_1$ and introducing the innerproduct

$$(h \oplus h_1)((v,v_1), (w,w_1)) = h(v,w) + h_1(v_1,w_1).$$

We shall denote this by $(V \oplus V_1, h \oplus h_1)$.

(1.1) <u>Definition</u>: <u>If</u> $W \subset V$ <u>is a subspace then</u> $W^\perp$ <u>is the subspace</u>

$$W^\perp = \{v \mid v \in V,\ h(v,W) = 0\}.$$

Notice that there is always a short exact sequence

$$0 \to W^{\perp} \to V \xrightarrow{r} \mathrm{Hom}_E(W,E) \to 0$$

in which $r(x)$ is the linear functional on W given by $\varphi_x(w) = h(w,x)$.

(1.2) Lemma: For $W \subset V$, $\dim W + \dim W^{\perp} = \dim V$ and $(W^{\perp})^{\perp} = W$.

Proof:

The statement involving dimensions follows from the exact sequence. It is clear that $W \subset (W^{\perp})^{\perp}$ and

$$\dim (W^{\perp})^{\perp} = \dim V - \dim W^{\perp} = \dim W$$

Hence the equality. ■

(1.3) Lemma: Suppose that W_1 and W_2 are subspaces of V, then

$$(W_1 + W_2)^{\perp} = W_1^{\perp} \cap W_2^{\perp}$$

$$(W_1 \cap W_2)^{\perp} = W_1^{\perp} + W_2^{\perp} .$$

Proof:

The first assertion is trivial. It implies, using (II.1.2),

$$(W_1^{\perp} + W_2^{\perp})^{\perp} = (W_1^{\perp})^{\perp} \cap (W_2^{\perp})^{\perp} = W_1 \cap W_2 .$$

Apply (II.1.2) again and the second equation follows. ■

(1.4) Definition: If $N \subset V$ is a subspace for which $N = N^{\perp}$ then N is a metabolizer for (V,h).

We note that $2 \dim N = \dim V$ for any metabolizer. We need an important lemma.

(1.5) Lemma: Suppose (V,h) has a metabolizer N. If $W \subset V$ is a subspace for which $W \subset W^{\perp}$ then $W + W^{\perp} \cap N$ is also a metabolizer.

Proof:

Appealing to (II.1.3),

$$(W + W^{\perp} \cap N)^{\perp} = W^{\perp} \cap (W^{\perp} \cap N)^{\perp} = W^{\perp} \cap (W + N).$$

Since $W \subset W^{\perp}$, $W^{\perp} \cap (W + N) = W + W^{\perp} \cap N$. ■

(1.6) Theorem: Suppose that (V,h) and (V_1,h_1) are Hermitian innerproduct spaces for which both (V_1,h_1) and $(V \oplus V_1, h \oplus h_1)$ admit metabolizers, then (V,h) also admits a metabolizer.

Proof:

Let $N_1 \subset V_1$ be a metabolizer for (V_1,h_1). Think of $N_1 \subset V \oplus V_1$ as $(0,N_1)$. Now with respect to $h \oplus h_1$,

$$(0,N_1)^{\perp} = V \oplus N_1 \supset (0,N_1).$$

Thus by (II.1.5) there is a metabolizer $\mathcal{N} \supset (0,N_1)$ for $(V \oplus V_1, h \oplus h_1)$.

Define a subspace $N \subset V$ by $v \in N$ if and only if there is a $v_1 \in V_1$ for which $(v,v_1) \in \mathcal{N}$. Since $(0,N_1) \subset \mathcal{N}$ it follows $v_1 \in N_1^{\perp} = N_1$ and thus $(v,0)$ is also contained in $\mathcal{N}$. In fact $\mathcal{N} = N \oplus N_1$. Clearly $N \subset N^{\perp}$ with respect to h. If $v \in N^{\perp}$ then $(v,0) \in \mathcal{N}^{\perp} = \mathcal{N}$ with respect to $h \oplus h_1$ and hence $v \in N$. So $N \subset V$ is a metabolizer for (V,h). ■

At last we are able to introduce the definition of Witt equivalence.

(1.7) Definition: Two Hermitian innerproduct spaces are Witt equivalent

$$(V,h) \sim_W (V_1,h_1)$$

<u>if and only if</u> $(V \oplus V_1, h \oplus (-h_1))$ <u>has a metabolizer</u>.

Obviously we need to prove this is an equivalence relation. First observe that the diagonal is a metabolizer for $(V \oplus V, h \oplus (-h))$. More generally if there is an isometry between (V,h) and (V_1,h_1) then the graph of that isometry is a metabolizer for $(V \oplus V_1, h \oplus (-h_1))$.

Next observe that a metabolizer for $(V \oplus V_1, h \oplus (-h_1))$ is also a metabolizer for $(V \oplus V_1, (-h) \oplus h_1)$, which is isometrically equivalent to $(V_1 \oplus V, h_1 \oplus (-h))$.

We need (II.1.6) for the transitivity. Thus $(V \oplus V_1, h \oplus (-h_1))$ and $(V_1 \oplus V_2, h_1 \oplus (-h_2))$ are both assumed to admit metabolizers, therefore so does

$$((V \oplus V_1) \oplus (V_1 \oplus V_2),\ ((h \oplus (-h_1))) \oplus (h_1 \oplus (-h_2))).$$

This sum can also be written as

$$(V \oplus (V_1 \oplus V_1) \oplus V_2,\ h \oplus ((-h_1) \oplus (h_1)) \oplus (-h_2)$$

which is isometrically equivalent to

$$((V \oplus V_2) \oplus (V_1 \oplus V_1),\ (h \oplus (-h_2)) \oplus (h_1 \oplus (-h_1)).$$

Now apply (II.1.6) to conclude that

$$(V,h) \sim (V_1,h_1). \quad \blacksquare$$

The Witt class of (V,h) will be denoted by $[V,h]$ and the collection of all such Witt classes by $H(E)$. Now $H(E)$ becomes an abelian group with respect to the operation

$$[V,h] + [V_1,h_1] = [V \oplus V_1,\ h \oplus h_1].$$

Of course $-[V,h] = [V,-h]$ and $[V,h] = 0$ if and only if (V,h) admits a metabolizer.

There is also a ring structure in $H(E)$ imposed by tensor product over E. Thus $(V \otimes V_1, h \otimes h_1)$ is defined to be the unique Hermitian innerproduct space for which $h \otimes h_1(v \otimes v_1, w \otimes w_1) = h(v,v_1)\cdot h(w,w_1)$. If $N \subset V$ is a metabolizer for (V,h) then $N \otimes V_1$ is a metabolizer for $(V \otimes V_1, h \otimes h_1)$. It is not difficult to verify that

$$[V,h]\cdot[V_1,h_1] = [V \otimes V_1, h \otimes h_1]$$

gives to $H(E)$ a commutative ring structure.

For $y \in F^*$ let $[y] \in H(E)$ be the Witt class of the innerproduct on E defined by $h(z_1,z_2) = yz_1\bar{z}_2$. Then $[1]$ is the multiplicative identity and $[y] \in H(E)^*$ for $y \in F^*$.

2. Anisotropic representatives

(2.1) **Definition:** <u>A Hermitian innerproduct space</u> (V,h) <u>is anisotropic if and only if</u> $W \cap W^\perp = \{0\}$ <u>for every subspace</u> W.

Note that if $W \cap W^\perp \neq \{0\}$ then setting $L = W \cap W^\perp$ we obtain a non-trivial subspace for which $L \subset L^\perp$. This will suggest an alternative equivalent definition of anisotropic.

(2.2) **Lemma:** <u>Let</u> (V,h) <u>be a Hermitian innerproduct space and let</u> W <u>be a subspace for which</u> $W \subset W^\perp$, <u>then a canonical innerproduct</u> $(W^\perp/W, \tilde{h})$ <u>is induced and</u> $[W^\perp/W, \tilde{h}] = [V,h]$.

<u>Proof:</u>

Let $\nu\colon W^\perp \to W^\perp/W$ be the quotient homomorphism. Define

$$\tilde{h}(\nu(v),\nu(v_1)) = h(v,v_1)$$

for v and v_1 in $W^\perp$. If w, w_1 lie in W then

$$h(v+w, v_1+w_1) = h(v,v_1) + h(v,w_1) + h(w,v_1) + h(w,w_1) = h(v,v_1)$$

so $\tilde{h}$ is well defined.

Next we want to show that $(W^\perp/W, \tilde{h})$ is an innerproduct space. We may as well consider a linear functional $\varphi\colon W^\perp \to E$ with $\varphi(W) = 0$. There is a $w \in V$ such that $h(v,w) = \varphi(v)$ for all $v \in W^\perp$. If $v \in W$ then $h(v,w) = \varphi(v) = 0$ and hence $w \in W^\perp$. If $w \in W^\perp$ and $h(v,w) = 0$ for all $v \in W^\perp$ then $w \in (W^\perp)^\perp = W$. Therefore

$$\text{Ad}\colon W^\perp/W \simeq \text{Hom}_E(W^\perp/W, E).$$

Thus $(W^\perp/W, \tilde{h})$ is an innerproduct. The metabolizer for

$$(V \oplus (W^\perp/W),\ h \oplus (-\tilde{h}))$$

is the graph of the quotient homomorphism $\nu\colon W^\perp \to W^\perp/W$. Let N denote this graph, then obviously $N \subset N^\perp$. Consider a pair $(v, \nu(w_0)) \in V \oplus (W^\perp/W)$ which lies in $N^\perp$. Then for every $w \in W^\perp$ we must have

$$h(v,w) = \tilde{h}(\nu(w_0), \nu(w)) = h(w_0, w).$$

Thus $h(v-w_0, w) = 0$ for all $w \in W^\perp$ and so $v - w_0 \in (W^\perp)^\perp = W \subset W^\perp$. This shows $v \in W^\perp$ and $\nu(v) = \nu(w_0)$. So we find $N = N^\perp$ as required. ■

The next result will return us to Witt's original approach to the Witt group.

(2.3) Theorem: Each Witt class contains an anisotropic representative which is unique up to an isometry.

Proof:

Suppose that (V,h) and (V_1,h_1) are both anisotropics and that $(V \oplus V_1,\ h \oplus (-h_1))$ has a metabolizer N. If we set $W = N \cap V$ then $W \subset W^\perp$ in (V,h)

and so by the anistropic assumption $W = \{0\}$. Similarly $N \cap V_1 = \{0\}$.

Let $K \subset V$ be the subspace of those vectors $v \in V$ such that $(v, v_1) \in N$ for some $v_1 \in V_1$. This v_1 is unique for if $(v, v_0) \in N$ also then $(0, v_1 - v_0) \in N \cap V_1$ and $v_1 - v_0 = 0$. So we have an isometry $L: K \to V_1$ for which N is the graph of L.

To finish we shall show $K^{\perp} = \{0\}$ so that $K = V$. Assume $w \in K^{\perp}$, then $(w, 0) \in N^{\perp} = N$ and hence $w = 0$. This completes uniqueness.

For existence consider a (V,h) and select a subspace maximal with respect to the property $W \subset W^{\perp}$. Then $(W^{\perp}/W, \tilde{h})$ is anisotropic and represents $[V,h]$. ■

(2.4) <u>Lemma</u>: <u>Let</u> (V,h) <u>be anisotropic</u>. <u>If</u> $W \subset V$ <u>is a subspace then the restriction of</u> h <u>to</u> W <u>is again an innerproduct space and</u> $V = W \oplus W^{\perp}$. ■

The proof is omitted as is the proof of

(2.5) <u>Lemma</u>: <u>If</u> (V,h) <u>is anisotropic then</u> V <u>has an orthogonal basis; that is, a basis</u> $e_1, \ldots, e_n$ <u>for which</u> $h(e_i, e_j) = 0$ <u>if</u> $i \neq j$. ■

Recall from (I.2) the group $H^2(C_2; E^*)$; that is, F^* modulo the subgroup of norms from E^*. There is a canonical embedding

$$1 \to H^2(C_2; E^*) \to H(E)^*.$$

Simply note that for y_1, y_2 in F^* the innerproducts on E given by

$$h_1(z, z_1) = y_1 z\bar{z}_1$$

$$h_2(z, z_1) = y_2 z\bar{z}_1$$

are anisotropics and are isometrically equivalent if and only if $c\ell(y_1) = c\ell(y_2)$ in $H^2(C_2; E^*)$. As a consequence of (II.2.5) we may say

(2.6) Lemma: The Witt group H(E) is additively generated by the image of

$$1 \to H^2(C_2;E^*) \to H(E)^*. \quad \blacksquare$$

3. Invariants for H(E)

Perhaps the most obvious invariant is

$$rk: H(E) \to F_2$$

which to each Witt class $[V,h]$ assigns $\dim_E V$ (mod 2). Since an innerproduct space with metabolizer has $\dim V = 2 \dim N$ it follows that if $(V,h) \sim_W (V_1,h_1)$ then $\dim V \equiv \dim V_1$ (mod 2). The kernel of this ring homomorphism is denoted by $J \subset H(E)$ and is called the fundamental ideal.

Our next invariant is the discriminant. It arises from the following considerations. Let (V,h) be a Hermitian innerproduct space. Choose a basis $e_1,\ldots,e_n$ for V, then $A = [h(e_i,e_j)]$ is a non-singular Hermitian matrix over E. If $f_1,\ldots,f_n$ is a second basis with associated matrix $B = [h(f_i,f_j)]$ then $A = MB\bar{M}^T$ for some $M \in GL(n,E)$. In particular $\det(A) = \det(M)\overline{\det(M)}\det(B)$. Hence associated to (V,h) there is a unique element

$$\det(V,h) \in H^2(C_2;E^*).$$

Obviously

$$\det(V \oplus V_1, h \oplus h_1) = \det(V,h)\det(V_1,h_1).$$

Unfortunately this is not a Witt class invariant. Suppose (V,h) admits a metabolizer N. Then $\dim V = 2k$ so choose a basis $e_1,\ldots,e_k$ for N and then select $e_{k+1},\ldots,e_{2k}$ in V so that

$$h(e_i, e_{j+k}) = \delta_{i,j} .$$

With respect to this basis the matrix of h will have the form

$$\left(\begin{array}{c|c} 0 & I_k \\ \hline I_k & X_k \end{array}\right)$$

and the determinant is $(-1)^k$.

For this reason the discriminant is defined by

$$\operatorname{dis}(V,h) = (-1)^{n(n-1)/2} \det(V,h)$$

in $H^2(C_2;E^*)$. Since

$$(-1)^{k(2k-1)}(-1)^k = (-1)^{2k^2}$$

it will follow that $\operatorname{dis}(V,h) = 1$ if (V,h) admits a metabolizer. Thus we now receive a Witt class invariant. Also observe the relation

$$(-1)^{(n+m)(n+m-1)/2} = (-1)^{mn}(-1)^{\frac{n(n-1)}{2}+\frac{m(m-1)}{2}}$$

This motivates the definition

(3.1) <u>Definition:</u> <u>The abelian group</u> Gp(E) <u>is the set</u>

$$H^2(C_2;E^*) \times F_2$$

with the group operation

$$(c\ell(y),\varepsilon)\cdot(c\ell(y_1),\varepsilon_1) = ((-1)^{\varepsilon\varepsilon_1}c\ell(yy_1),\ e+\varepsilon_1).$$

As a corollary to this definition

(3.2) Lemma: *There is an epimorphism*

$$H(E) \to Gp(E) \to 1$$

which to $[V,h]$ *assigns* $(\text{dis}[V,h],\ \text{rk}[V,h])$. ■

We point out that this is a homomorphism of the additive structure of $H(E)$. Also notice that

$$\text{dis}|J \to H^2(C_2;E^*) \to 1$$

is a homomorphism. We want to expand (II.3.2) into a short exact sequence.

First of all any element in J can be thought of as a sum of terms of the form $[y_1]+[y_2]$. Consequently an element of the ideal J^2 is a sum of products of the form

$$([y_1]+[y_2])([y_3]+[y_4])$$
$$= [y_1y_3]+[y_1y_4]+[y_2y_3]+[y_2y_4].$$

Thus we may say

(3.3) Lemma: *The square of the fundamental ideal lies in the kernel of*

$$\text{dis}|J \to H^2(C_2;E^*).$$ ■

The next lemma is from [M-H, p. 77].

(3.4) **Lemma**: *Modulo* J^2, *every element* $X \in J$ *is equal to* $[\text{dis } X] - [1]$.

Proof:

From

$$([y_1] + [1])([y_1] + [1]) = [y_1 y_2] + [y_1] + [y_2] + [1]$$

we learn

$$[y_1] + [y_2] = [-y_1 y_2] - [1] \pmod{J^2}.$$

In particular

$$[y] + [1] = [-y] - [1] \pmod{J^2}.$$

Next from

$$([1] + [1])([1] + [1]) = [1] + [1] + [1] + [1]$$

we conclude

$$-[1] - [1] - [1] = [1] \pmod{J^2}.$$

Now any element of J can be written as $[y_1] + \ldots + [y_{2r}]$, so by inductively applying the foregoing relations we can conclude that for $X \in J$

$$X = [y] - [1] \pmod{J^2}$$

for some $c\ell(y) \in H^2(C_2;E^*)$. Since discriminant is 1 on J^2 it follows

$$\text{dis } X = \text{dis}([y] - [1]) = c\ell(y). \quad \blacksquare$$

(3.5) **Lemma**: *Let* (V,h) *be an innerproduct space with* $\dim V = 2$, *then* $[V,h] = 0$ *if and only if* $\text{dis}[V,h] = 1 \in H^2(C_2;E^*)$.

Proof:

A moment's reflection shows that either $[V,h]=0$ or (V,h) is anisotropic. Let us assume (V,h) is anisotropic and prove it has non-trivial discriminant. Use (II.2.5) to choose an orthogonal basis e_1, e_2, so that any vector v can be Written

$$v = z_1e_1 + z_2e_2$$

and

$$h(v,v) = z_1\bar{z}_1h(e_1,e_1) + z_2\bar{z}_2h(e_2,e_2).$$

Write $y_1 = h(e_1,e_1)$, $y_2 = h(e_2,e_2)$. The discriminant is $c\ell(-y_1y_2) \in H^1(C_2;E^*)$. If this were trivial then $\xi\bar{\xi} = -y_1y_2$ and

$$y_1 + (\xi y_2^{-1})(\overline{\xi y_2^{-1}})y_2 = 0$$

which contradicts the anistropic assumption. ■

(3.6) Theorem: There is a short exact sequence

$$0 \to J^2 \to H(E) \to Gp(E) \to 1.$$

Proof:

We showed J^2 was in the kernel of dis and for $X \in J$

$$X = [\text{dis } X] - [1] \pmod{J^2}.$$

Finally $[\text{dis } X] - [1] = 0$ if and only if $\text{dis } X = 1 \in H^2(C_2;E^*)$. ■

As a corollary

(3.7) Pfister: Discriminent induces an isomorphism

$$\text{dis } J/J^2 \simeq H^1(C_2;E^*). \blacksquare$$

Up to this point it will be observed that our definitions and results only require that E be a field with an involution and that F be the field of fixed elements under this involution. There is no restriction on the characteristic of the field. In fact we could take the involution to be the identity in which case $E = F$ and we receive the Witt ring $W(E)$ of symmetric innerproduct spaces over E. In this case $E^*/E^{**} = H^2(C_2;E^*)$. Let us finish this section with

(3.8) Theorem: Let E be a finite field. If the involution is not the identity, then $\text{rk}: H(E) \simeq F_2$. If the involution is the identity then $W(E) \simeq Gp(E)$.

Proof:

When the involution is non-trivial the norm $N: E^* \to F^*$ is an epimorphism. As a consequence of (II.3.5) then there can be no 2-dimensional anisotropics, thus $J = \{0\}$ and we apply (II.3.6). If the involution is trivial then we must invoke the result that an anisotropic innerproduct space over a finite field has rank no greater than 2 [M-H, p. 21]. As a consequence $J^2 = \{0\}$. ■

If E is a finite field of characteristic 2 then every element is a square and $\text{rk}: W(E) \simeq F_2$. If E had odd characteristic then $W(E) \simeq (Z/2Z)[t]/(1+t^2)$ or $Z/4Z$ depending on whether or not -1 is a square in E.

4. Algebraic number fields

In this section we shall assume that E/F is a degree 2 extension of an algebraic number field. The involution is non-trivial. In this case Landherr [La] has shown how the isometry class of a Hermitian innerproduct space (V,h) is determined. The first two invariants are $\dim_E V$, called the rank of (V,h), and the discriminant of (V,h) in $H^2(C_2;E^*)$. The final invariant is called the <u>total signature</u>. This is discussed in [M-H, p. 61-62] in connection with $W(\cdot)$, however we shall take up the Hermitian cases.

(4.1) <u>Lemma</u>: <u>For any Hermitian innerproduct space over</u> E <u>there is an orthogonal basis</u>.

<u>Proof</u>:

First we assert there is a vector $v \in V$ with $h(v,v) \neq 0$. If this were not so then for any pair v,w in V, $0 = h(v+w,v+w) = h(v,v) + h(v,w) + h(w,v) + h(w,w)$ would imply

$$\overline{h(v,w)} = -h(v,w).$$

But then

$$\overline{h(v,w)} = \overline{h(\sqrt{\sigma}\,(\sqrt{\sigma})^{-1}v,w)} = \overline{\sqrt{\sigma}\,h((\sqrt{\sigma})^{-1}v,w)}$$
$$= -\sqrt{\sigma}\,(-h((\sqrt{\sigma})^{-1}v,w) = h(v,w)$$

would (now) show $h(v,w) \equiv 0$.

Thus we pick an e_1 in V with $h(e_1,e_1) \neq 0$. Let $W \subset V$ be the subspace of all vectors w with $h(w,e_1) = 0$. This is just the orthogonal complement of the subspace generated by e_1. This yields an orthogonal direct sum splitting since any vector in V can be expressed as

$$v = \frac{h(v,e_1)}{h(e_1,e_1)}\,e_1 + \quad v - \frac{h(v,e_1)}{h(e_1,e_1)}\,e_1$$

Thus the restriction of h to W is still an innerproduct. By induction over dimension we can assume W already has an orthogonal basis $e_2, \ldots, e_n$. A note of caution. This lemma does not extend to fields of characteristic 2, which is why we gave (II.2.5) separately.

Now suppose P_∞ is a real infinite prime of F that ramifies. Recall then that every norm from E^* is positive for P_∞.

(4.2) Lemma: Let P_∞ be a ramified real infinite prime and let (V, h) be an inner product space over E. Then there is an orthogonal direct sum decomposition $V = X^+ \oplus X^-$ in which $h(v,v) > 0$ at P_∞ for all $0 \neq v \in X^+$ and $h(v,v) < 0$ for all $0 \neq v \in X^-$. Furthermore $\dim X^+$ and $\dim X^-$ are independent of the choice of such a decomposition.

Proof:

Let $e_1, \ldots, e_n$ be an orthogonal basis. Take X^+ to be the subspace generated by all e_j with $h(e_j, e_j) > 0$ and X^- to be the subspace generated by all e_i with $h(e_i, e_i) < 0$. This shows existence. Let $W \subset V$ be any subspace on which h is positive definite with respect to P_∞. Then $W \cap X^- = \{0\}$ and hence $\dim W \leq \dim V - \dim X^- = \dim X^+$. Therefore $\dim X^+$ is the largest dimension possible for a subspace on which h is positive definite with respect to P_∞. In the same way $\dim X^-$ is seen also to be an invariant of (V, h). ■

The signature of (V, h) at P_∞ is defined to be

$$\operatorname{sgn}_{P_\infty}(V,h) = \dim X^+ - \dim X^-.$$

This can vary from one ramified real infinite prime to the next. This is easily seen to be additive with respect to orthogonal sums and multiplicative with respect to tensor products Actually $\operatorname{sgn}_{P_\infty}(V,h)$ is a Witt class invariant. Suppose

$V = X^+ \oplus X^-$, then $N \cap X^+ = \{0\}$ and $N \cap X^- = \{0\}$ so $\dim N \leq \min(\dim X^+, \dim X^-)$. On the other hand $2 \dim N = \dim V = \dim X^+ + \dim X^-$ so $\dim X^+ = \dim X^-$. Thus we receive a ring homomorphism.

(4.3) Convention: *If at least one real infinite prime in F ramifies then E/F is said to have signatures. Otherwise E/F has no signatures.*

When E has signatures let Z(E) be the direct sum of the ring of rational integers Z with itself, one copy for each ramified real infinite prime. The *total signature* $\mathrm{Sgn}(V,h) \in Z(E)$ is defined in the obvious manner.

(4.4) Landherr: *The isometry class of (V,h) is uniquely determined by*

1) $\mathrm{rank}(V,h) = \dim V$

2) $\mathrm{dis}(V,h) \in H^2(C_2;E^*)$

3) and, if E/F has signatures,

$$\mathrm{Sgn}(V,h) \in Z(E). \quad \blacksquare$$

To translate this into the structure of H(E) note that if $\dim(V,h) = 2k$, $\mathrm{dis}(V,h) = 1 \in H^2(C_2;E^*)$ and $\mathrm{Sgn}(V,h) = 0$ (if there are signatures) then by (II.4.4), (V,h) is isometrically equivalent to $\begin{array}{c|c} I_k & 0 \\ \hline 0 & -I_k \end{array}$ and so $(V,h) \sim_W 0$.

(4.5) Corollary: *If E/F has no signatures then*

$$H(E) \simeq Gp(E)$$

and H(E) contains an element of order 4 if and only if $c\ell(-1) \neq 1$ in $H^2(C_2;E^*)$. $\blacksquare$

Assume then that signatures are present. Note that for each real ramified P_∞, $sgn_{P_\infty}(V,h) \equiv \dim V \pmod 2$. Thus all signatures are congruent mod 2. If $y \in F^*$ then $sgn_{P_\infty}[y] = (y,\sigma)_{P_\infty}$ for every ramified real infinite prime. Using realization of Hilbert symbols we may summarize as

(4.6) <u>Corollary</u>: *If E has signatures then the composition*

$$H^2(C_2;E^*) \to H(E)^* \to Z(E)^*$$

is an epimorphism,

$$Sgn|J \to 2\, Z(E)$$

is an epimorphism, and

$$Sgn|J^2 \simeq 4\, Z(E). \blacksquare$$

The point of course is that by the time we get to J^2 the only invariant left is total signature.

(4.7) <u>Corollary</u>: *If E/F has signatures there is*

$$0 \to 4\, Z(E) \to H(E) \to Gp(E) \to 1.$$

The torsion ideal in $H(E)$ is the kernel of $Sgn|J$. Every torsion class has order 2. The product of any two torsion classes in 0. ■

We saw a relation between rank mod 2 and signatures. There is still another relation

(4.8) <u>Lemma</u>: *If $[V,h] \in J$ and E/F has signatures then for each ramified real infinite prime*:

$$(\text{dis}[V,h],\sigma)_{P_\infty} = (-1)^{(\text{sgn}_{P_\infty}[V,h])/2}$$

We should also mention explicitly the signature realization theorem.

(4.9) Theorem: Let E/F have signatures. To each finite prime P assign $\varepsilon_P \varepsilon Z^*$ so that $\varepsilon_P = 1$ for all P split and $\varepsilon_P = 1$ for almost all primes. To each ramified real ramified infinite prime P_∞ assign an even integer $2n_{P_\infty}$. Then there is $X \in J \subset H(E)$ with

$$(\text{dis } X,\sigma)_P = \varepsilon_P$$

at all finite primes and

$$\text{Sgn}_{P_\infty} X = 2n_{P_\infty}$$

for all ramified real infinite primes if and only if

$$\prod_{\text{finite } P} \varepsilon_P = (-1)^{n_{P_{\infty_1}} + \dots + n_{P_{\infty_r}}}.$$

Proof:

Begin with (II.4.8). For X in J we know that modulo J^2, $X = [\text{dis } X] - [1]$. We also know discriminant is trivial on J^2 and $\text{Sgn}|J^2 \simeq 4\,Z(E)$. Hence it is enough to verify the lemma for $[y] - [1]$. This has discriminant $c\ell(y) \in H^2(C_2;E^*)$. Since at each ramified real P_∞

$$\text{sgn}_{P_\infty}([y]-[1]) = (y,\sigma)_{P_\infty} - 1$$

we can say $(y,\sigma)_{P_\infty} = (-1)^{\text{sgn}_{P_\infty}([y]-[1])/2}$ which is what we want. For (II.4.9) we remark that it is enough to consider that the assigned integers $n_{P_{\infty_1}}, \ldots, n_{P_{\infty_r}}$ are all in the set $\{0,-1\}$. You already know $\mathrm{dis}|J^2 \simeq 4\,Z(E)$. So this is reduced to Hilbert symbol realization to obtain the correct element $[y]-[1]$. ■

To close this section we may as well go on to determine the group of units $H(E)^*$. Continue to assume E/F has signatures. If $X \in H(E)^*$ then $\mathrm{Sgn}(X) \in Z(E)^*$. By (II.4.6) there is a $y \in F^*$ with $\mathrm{Sgn}[y] = \mathrm{Sgn}(X)$. Thus $[y] - X \in J$ is a torsion element.

(4.10) <u>Lemma</u>: <u>If E/F has signatures then $H(E)^*$ is the group of elements of the form $[y]+\tau$, $y \in F^*$, $\tau \in J$ a torsion class. If E/F has no signatures, then $H(E)^*$ consists of the elements $[1]+\tau$, $\tau \in J$.</u> ■

In either case all torsion classes in J are 2-nilpotent so $[y]+\tau$ has inverse $[y^{-1}] - [y^{-2}]\tau$. If no signatures are present then J is exactly the ideal of non-units in H(E).

III Torsion Forms

1. Torsion O_E-modules

In this chapter M is a finitely generated torsion O_E-module. Since M is finitely generated there is an ideal $\mathrm{Ann}(M) \subset O_E$ defined by

$$\mathrm{Ann}(M) = \{\lambda \mid \lambda \in O_E, \lambda x = 0, \text{ all } x \in M\}.$$

This annihilator ideal has a unique factorization into a product of powers of distinct prime ideals. Suppose $\mathscr{P} \mid \mathrm{Ann}(M)$ with exponent $i > 0$. Then the $\mathscr{P}$-primary torsion submodule of M is defined to be

$$M_{\mathscr{P}} = \{x \mid x \in M, \alpha x = 0, \text{ all } \alpha \in \mathscr{P}^i\}.$$

(1.1) Lemma: The finitely generated torsion O_E-module M is the direct sum of its $\mathscr{P}$-primary submodules taken over all $\mathscr{P}$ dividing $\mathrm{Ann}(M)$.

Proof:

This is trivial if $\mathrm{Ann}(M)$ is a power of a single prime. Thus write $\mathrm{Ann}(M) = \mathscr{P}^i K$ where $K \subset O_E$ is an ideal realtively prime to $\mathscr{P}$. Let now

$$M_K = \{x \mid x \in M, \beta x = 0, \text{ all } \beta \in K\}.$$

Now the claim is that M is the direct sum of $M_{\mathscr{P}}$ with M_K. Surely we can find $\alpha \in \mathscr{P}^i$, $\beta \in K$ with $\alpha + \beta = 1$. Thus if $x \in M_{\mathscr{P}} \cap M_K$, $x = \alpha x + \beta x = 0$. Then for $x \in M$ write $x = \alpha x + \beta x$. Since $\alpha K \subset \mathscr{P}^i \cap K$, it follows $\alpha x \in M_K$ and similarly $\beta x \in M_{\mathscr{P}}$.

We should also point out the relation $\mathscr{P}^i M = M_K$, for if $x \in M_K$ then $x = \alpha x$. Similarly $KM = M_{\mathscr{P}}$. Thus $\mathrm{Ann}(M_K) = K$ and inductively we may suppose M_K has been decomposed into a sum of its primary submodules. ■

(1.2) Lemma: If $M_{\mathscr{P}}$ is the $\mathscr{P}$-primary submodule of M then for any $\lambda \in O_E \setminus \mathscr{P}$ multiplication by λ is on O_E-module automorphism of $M_{\mathscr{P}}$.

Proof:

The ideals λO_E and $\mathscr{P}^i$ are relatively prime so $\lambda c + \alpha = 1$ for some $c \in O_E$ and $\alpha \in \mathscr{P}^i$. Thus for $x \in M_{\mathscr{P}}$, $x = \lambda(cx) = c(\lambda x)$.

Now this implies $M_{\mathscr{P}}$ naturally receives the structure of an $O_E(\mathscr{P})$-module. (Parenthetically, M is then a module over the semi-local ring $\cap O_E(\mathscr{P})$ taken over all $\mathscr{P}|\text{Ann}(M)$. A semi-local ring here is a P.I.D.) But $O_E(\mathscr{P})$ is a valuation ring and so we know the structure of a finitely generated torsion $O_E(\mathscr{P})$-module. Let $m(\mathscr{P}) \subset O_E(\mathscr{P})$ be the unique maximal ideal in $O_E(\mathscr{P})$, then for $j > 0$, $O_E(\mathscr{P})/(m(\mathscr{P}))^j$ is a torsion cyclic module and $M_{\mathscr{P}}$ can be written as a direct sum of torsion cyclic $O_E(\mathscr{P})$-modules with $j \leq i$ where $i = \text{ord}_{\mathscr{P}}(\text{Ann}(M))$. In fact the obvious numerical invariants will characterize a finitely generated torsion $O_E(\mathscr{P})$-module up to isomorphism.

2. The quotient E/K

Allow $K = \bar{K} \subset E$ to be any conjugation invariant fractional O_E-ideal. We regard E/K as an O_E-ideal. It is a torsion O_E-module, but not finitely generated. It is O_E-injective and it inherits an involution. We particularly draw the reader's attention to the observation that $z \in K$ if and only if $\text{ord}_{\mathscr{P}} z \geq \text{ord}_{\mathscr{P}} K$ for all $\mathscr{P} \subset O_E$.

If M is a finitely generated torsion O_E-module we set

$$M^* = \text{Hom}_{O_E}(M, E/K)$$

and define the O_E-module structure on M^* by $(\lambda\varphi)(x) = \varphi(\bar{\lambda}x)$ for $\lambda \in O_E$, $\varphi: M \to E/K$ and $x \in M$.

For each prime $\mathcal{P}$ there is the localization of K denoted by $K_{(\mathcal{P})}$. This is $K \cdot O_E(\mathcal{P})$. Alternatively it is all elements of E which can be written as k/λ with $k \in K$ and $\lambda \in O_E \setminus \mathcal{P}$. Finally, it is all $z \in E$ with $\operatorname{ord}_{\mathcal{P}} z \geq \operatorname{ord}_{\mathcal{P}} K$. In any case $K_{(\mathcal{P})}$ is a fractional $O_E(\mathcal{P})$ ideal. The fractional $O_E(\mathcal{P})$ ideals form an infinite cyclic group generated by $m(\mathcal{P})$. Thus $K_{(\mathcal{P})} = (m(\mathcal{P}))^{\nu_{\mathcal{P}}(K)}$. So for almost all primes $K_{(\mathcal{P})} = O_E(\mathcal{P})$. Furthermore

$$K = \bigcap_{\mathcal{P}} K_{(\mathcal{P})} .$$

(2.1) Lemma: There is a natural isomorphism of O_E-modules

$$E/K \simeq \sum E/K_{(\mathcal{P})} .$$

Proof:

We had just noted this would be a monomorphism. To see it is an epimorphism we must show the following. Given primes $\mathcal{P}_1, \ldots, \mathcal{P}_r$ and elements $z_1, \ldots, z_r$ there is a $z \in Z$ such that

$$\operatorname{ord}_{\mathcal{P}} z \geq \operatorname{ord}_{\mathcal{P}} K \qquad \mathcal{P} \notin \{\mathcal{P}_1, \ldots, \mathcal{P}_r\}$$

$$\operatorname{ord}_{\mathcal{P}_i} (z - z_i) \geq \operatorname{ord}_{\mathcal{P}_i} K, \quad \text{each } \mathcal{P}_i .$$

Since $\operatorname{ord}_{\mathcal{P}} K = 0$ for almost all primes this follows by the Strong Approximation Theorem [O'M, p. 42]. ■

Obviously we want to think of M^* as $\sum \operatorname{Hom}_{O_E(\mathcal{P})}(M_{\mathcal{P}}, E/K_{(\mathcal{P})})$ with sum over all $\mathcal{P} \mid \operatorname{Ann}(M)$. One point should be carefully noted. We must regard $\operatorname{Hom}_{O_E(\mathcal{P})}(M_{\mathcal{P}}, E/K_{(\mathcal{P})})$ as an $O_E(\overline{\mathcal{P}})$-module because in general $\mathcal{P}$ is not equal to $\overline{\mathcal{P}}$. With this in mind we can say that as O_E-modules

$$M^* \simeq \sum \mathrm{Hom}_{O_E(\mathscr{P})}(M_{\mathscr{P}}, E/K_{(\mathscr{P})}).$$

We may analyse $\mathrm{Hom}_{O_E(\mathscr{P})}(M_{\mathscr{P}}, E/K_{(\mathscr{P})})$ in terms of torsion cyclic modules. Choose a local uniformizer $\pi \in O_E(\mathscr{P})$ and a torsion cyclic module $O_E(\mathscr{P})/(\pi^r)$, $r > 0$. The observation is that there is an $O_E(\bar{\mathscr{P}})$-module isomorphism

$$O_E(\bar{\mathscr{P}})/(\bar{\pi}^r) \simeq \mathrm{Hom}_{O_E(\mathscr{P})}(O_E(\mathscr{P})/(\pi^r), E/K_{(\mathscr{P})}).$$

It is specifically

$$g: O_E(\mathscr{P})/(\pi^r) \to E/K_{(\mathscr{P})}$$

given by

$$g(\lambda) = (\lambda\pi^{j-r}) \in E/K_{(\mathscr{P})}, \quad \lambda \in O_E(\mathscr{P}),$$

where $j = \mathrm{ord}_{\mathscr{P}} K$ which generates $\mathrm{Hom}_{O_E(\mathscr{P})}(O_E(\mathscr{P})/(\pi^r), E/K_{(\mathscr{P})})$ as a $O_E(\bar{\mathscr{P}})/(\bar{\pi}^r)$-module.

We come to a central question. If M is a finitely generated torsion O_E-module then M^* is defined with its O_E-module structure and hence there is $(M^*)^*$.

As we would expect, there is an O_E-module homomorphism $M \to M^{**}$. For $y \in M$, $\varphi \in M^*$ let

$$y^{**}(\varphi) = \overline{\varphi(y)}.$$

First we have to show y^{**} is O_E-linear. Clearly it is additive. For $\lambda \in O_E$

$$y^{**}(\lambda\varphi) = \overline{(\lambda\varphi)(y)} = \overline{\varphi(\bar{\lambda}y)}$$

$$= \lambda\overline{\varphi(y)} = \lambda(y^{**}(\varphi)).$$

Next we must show $y \to y^{**}$ is an O_E-module homomorphism of M into M^{**}. First, for $\lambda \in O_E$

$$(\lambda y)^{**}(\varphi) = \overline{\varphi(\lambda y)} = \bar{\lambda}\ \overline{\varphi(y)} .$$

Next observe that

$$(\lambda y^{**})(\varphi) = y^{**}(\bar{\lambda}\varphi) = \overline{(\bar{\lambda}\varphi)(y)} = \overline{\varphi(\lambda y)} = \bar{\lambda}\ \overline{\varphi(y)}.$$

So we find $(\lambda y)^{**} = \lambda y^{**}$ as required.

Now suppose we look at

$$\mathrm{Hom}_{O_E(\mathscr{P})}(M_{\mathscr{P}}, E/K_{(\mathscr{P})}).$$

This is an $O_E(\bar{\mathscr{P}})$-module, and

$$(M_{\mathscr{P}}^{*})^{*} = \mathrm{Hom}_{O_E(\bar{\mathscr{P}})}(\mathrm{Hom}_{O_E(\mathscr{P})}(M_{\mathscr{P}}, E/K_{(\mathscr{P})}), E/K_{(\bar{\mathscr{P}})})$$

is again an $O_E(\mathscr{P})$-module. Furthermore, the definition $M_{\mathscr{P}} \to (M_{\mathscr{P}}^{*})^{*}$ will work as above. That is, if $y \in M_{\mathscr{P}}$ and $\varphi: M_{\mathscr{P}} \to E/K_{(\mathscr{P})}$, then $\overline{\varphi(y)} \in E/K_{(\bar{\mathscr{P}})}$. The correspondence $y^{**}(\varphi) = \overline{\varphi(y)}$ is $O_E(\bar{\mathscr{P}})$ linear with y fixed and $y \to y^{**}$ is $O_E(\mathscr{P})$-linear.

(2.2) Theorem: If M is a finitely generated torsion O_E-module then $M \simeq M^{**}$.

Proof:

Obviously we are only going to consider a torsion cyclic $O_E(\mathscr{P})$-module. Using a local uniformizer we saw

$$\mathrm{Hom}_{O_E(\mathscr{P})}(O_E(\mathscr{P})/(\pi^r), E/K_{(\mathscr{P})})$$

is free as an $O_E(\bar{\mathscr{P}})/(\bar{\pi}^r)$-module on the generator $g(\lambda) = (\lambda\pi^{j-r}) \in E/K_{(\mathscr{P})}$. Now for $\lambda \in O_E(\mathscr{P})/(\pi^r)$ the effect of λ^{**} on $g \in \mathrm{Hom}_{O_E(\mathscr{P})}(O_E(\mathscr{P})/(\pi^r), E/K_{(\mathscr{P})})$ is $\lambda^{**}(g) = \overline{g(\lambda)} = (\overline{\lambda\pi}^{j-r}) \in E/K_{(\bar{\mathscr{P}})}$. Now if $O_E(\bar{\mathscr{P}})/(\bar{\pi}^r)$ is thought of as an $O_E(\mathscr{P})$-module then $\lambda \to \bar{\lambda}$ is an $O_E(\mathscr{P})$-module isomorphism

$$O_E(\mathscr{P})/(\pi^r) \simeq O_E(\bar{\mathscr{P}})/(\bar{\pi}^r).$$

Hence (III.2.2) follows. ■

3. Torsion innerproducts

We shall use a somewhat more general definition than might be expected. This has advantages of unification, it fits well with Wall's definitions and it has some further applications of its own [St]. We fix a fractional O_E-ideal K with $K = \bar{K}$ and a unit $u \in O_E^*$ with $u\bar{u} = 1$. By a <u>torsion</u> u-<u>innerproduct</u> we shall mean a pair (M, b) where

1) M is a finitely generated torsion O_E-module

2) $b: M \times M \to E/K$ is a biadditive function such that

$$\lambda b(x,y) = b(\lambda x, y) = b(x, \bar{\lambda} y)$$

for all $\lambda \in O_E$ and

$$\overline{ub(x,y)} = b(y,x).$$

3) Ad: $M \to M^*$ given by $y \to \varphi_y(x) = b(x,y)$ is an isomorphism of O_E-modules.

We want to make up a Witt theory so we shall need a series of lemmas and definitions analogous to those of II.1 and II.2.

(3.1) Definition: If $L \subset M$ is a submodule then

$$L^{\perp} = \{y \mid y \in M,\ b(y,L) = 0\}.$$

Notice here that $b(L,y) = u\overline{b(y,L)}$ so that since u is a unit we can say that $y \in L^{\perp}$ if and only if $b(L,y) = 0$.

(3.2) Lemma: For any submodule $L \subset M$ there is a short exact sequence

$$0 \to L^{\perp} \to M \xrightarrow{r} \mathrm{Hom}_{O_E}(L, E/K) \to 0$$

of O_E-modules.

Proof:

The homomorphism r sends $y \in M$ to $\varphi_y(x) = b(x,y) \in E/K$ for all $x \in L$. To see that this is an epimorphism we must recognize E/K is a divisible O_E-module and is therefore injective. So any homomorphism $\psi\colon L \to E/K$ has an extension $\varphi\colon M \to E/K$ which can be represented as $b(\cdot,y)$. ■

Let $L^* = \mathrm{Hom}_{O_E}(L, E/K)$. Appealing again to the injectivity of E/K we find that there must be a dual short exact sequence

$$0 \to (L^*)^* \xrightarrow{r^*} M^* \xrightarrow{i^*} (L^{\perp})^* \to 0$$

which must be closely related to

$$0 \to (L^{\perp})^{\perp} \xrightarrow{j} M \xrightarrow{r'} (L^{\perp})^* \to 0.$$

Indeed, using $\mathrm{ad}\colon M \simeq M^*$ it is immediate that we have a commutative diagram:

$$\begin{array}{ccc} M^* & \xrightarrow{i^*} & (L^{\perp})^* \to 0 \\ \text{ad}\uparrow\simeq & & \text{id}\uparrow\simeq \\ M & \xrightarrow{r'} & (L^{\perp})^* \to 0. \end{array}$$

Therefore there is an O_E-module isomorphism

$$q\colon (L^{\perp})^{\perp} \simeq (L^*)^*$$

for which

$$\begin{array}{ccccc} 0 \to (L^*)^* & \xrightarrow{r^*} & M^* & \xrightarrow{i^*} & (L^{\perp})^* \to 0 \\ \simeq\uparrow q & & \simeq\uparrow\text{ad} & & \simeq\uparrow\text{id} \\ 0 \to (L^{\perp})^{\perp} & \xrightarrow{j} & M & \xrightarrow{r'} & (L^{\perp})^{\perp} \to 0 \end{array}$$

is a commutative diagram. Obviously $L \subset (L^{\perp})^{\perp}$ so let us see what $q(y) \in (L^*)^*$ is for a $y \in L$. Now with $x \in M$ we have

$$r^*(q(y))(x) = q(y)(\varphi_x|L \to E/K) = \varphi_y(x).$$

Thus $b(x,y) = q(y)(\varphi_x|L)$. But $b(x,y) = u\overline{b(y,x)} = uy^{**}(\varphi_x)$. Obviously

$$q|L \simeq \dot{L}^{**}$$

also. This shows

(3.3) Lemma: If $L \subset M$ is a submodule of a torsion u-innerproduct (M,b) then $L = (L^{\perp})^{\perp}$. ■

Actually this is all we really need. Let us introduce

(3.4) Definition: If (M,b) is a torsion u-innerproduct, then a metabolizer is an O_E-submodule $N \subset M$ for which $N^{\perp} = N$. Furthermore $(M,b) \sim_W (M_1,b_1)$ if and only if $(M \oplus M_1, b \oplus (-b_1))$ admits a metabolizer.

We shall then suggest

(3.5) Exercise: Prove that $(M,b) \sim_W (M_1,b_1)$ is an equivalence relation on torsion u-innerproducts. Define the Witt group $H_u(E/K)$.

We wish to eliminate from further consideration those prime ideals for which $\mathcal{P} \neq \overline{\mathcal{P}}$.

(3.6) Lemma: Let (M,b) be a torsion u-innerproduct. Let $\mathcal{P}$, $\mathcal{P}_1$ be two primes dividing $\mathrm{Ann}(M)$. If $\mathcal{P}_1 \neq \overline{\mathcal{P}}$ then for $x \in M_{\mathcal{P}}$, $y \in M_{\mathcal{P}_1}$,

$$b(x,y) = 0.$$

Proof:

Let $i \geq 1$ be $\mathrm{ord}_{\mathcal{P}} \mathrm{Ann}(M)$ and let $j \geq 1$ be $\mathrm{ord}_{\mathcal{P}_1} \mathrm{Ann}(M)$. Since $\mathcal{P}_1 \neq \overline{\mathcal{P}}$ there is $\alpha \in \mathcal{P}_1^j$ and $\beta \in \overline{\mathcal{P}}^i$ such that $\alpha + \beta = 1$. Then $b(x,y) = b(x, \alpha y + \beta y) = b(x, \beta y) = b(\overline{\beta} x, y) = 0$. ■

Thus if $\mathcal{P} = \overline{\mathcal{P}}$ then $M_{\mathcal{P}}$ is an orthogonal direct summand of M and by restriction, $(M_{\mathcal{P}}, b)$ is a torsion u-innerproduct over $O_E(\mathcal{P})$ with values in $E/K_{(\mathcal{P})}$. However, if $\mathcal{P} \neq \overline{\mathcal{P}}$ then by restriction $(M_{\mathcal{P}} \oplus M_{\overline{\mathcal{P}}}, b)$ must be a torsion u-innerproduct, but then $M_{\mathcal{P}}$ is obviously a metabolizer.

(3.7) Theorem: There is a natural isomorphism

$$H_u(E/K) \simeq \sum_{\mathcal{P}=\bar{\mathcal{P}}} H_u(E/K_{(\mathcal{P})}). \quad \blacksquare$$

Now we have localized the computation and we wish to find $H_u(E/K_{(\mathcal{P})})$ in terms of the residue field $O_E(\mathcal{P})/m(\mathcal{P}) = E_{\mathcal{P}}$.

(3.8) Definition: Let (M,b) be a finitely generated torsion $O_E(\mathcal{P})$-module with a torsion u-innerproduct having values in $E/K_{(\mathcal{P})}$. Then (M,b) is anistropic if and only if $L \cap L^{\perp} = \{0\}$ for every submodule $L \subset M$.

Following (II.2.3) show

(3.9) Exercise: Every Witt class in $H_u(E/K_{(\mathcal{P})})$ contains an anisotropic representative unique up to isometry.

Suppose (M,b) is an anisotropic. We assert $\mathrm{Ann}(M) = m(\mathcal{P})$. If this were not the case, then we could find $\lambda \in O_E(\mathcal{P})$ with $\lambda \notin \mathrm{Ann}(M)$ but $\lambda^2 \in \mathrm{Ann}(M)$ together with $x \in M$ so that $\lambda x \neq 0$. Now $\bar{\lambda} = \lambda v$ for some $v \in O_E(\mathcal{P})^*$ therefore

$$b(\lambda x,\lambda x) = b(\lambda\bar{\lambda}x,x) = vb(\lambda^2 x,x) = 0$$

and this violates the anisotropic assumption.

Hence we have shown that for an anisotropic $(M_{\mathcal{P}}, b)$ the module is naturally a vector space over the residue field $E_{\mathcal{P}}$. There is as before an orthogonal basis $e_1, \ldots, e_n$ with $b(e_i,e_j) = 0$ if $i \neq j$. For an anisotropic $(M_{\mathcal{P}}, b)$ the values $b(x,y)$ all lie in $\Gamma_{\mathcal{P}} \subset E/K_{(\mathcal{P})}$, the unique submodule with $\mathrm{Ann}(\Gamma_{\mathcal{P}}) = m(\mathcal{P})$. This $\Gamma_{\mathcal{P}}$ is a 1-dimensional vector space over $E_{\mathcal{P}}$. Now we shall introduce a concept which will be used throughout the rest of these notes.

(3.10) Definition: A localizer for K at $\mathcal{P}$ is an element $\rho \in E^*$ for which $\mathrm{ord}_{\mathcal{P}}\, \rho = \mathrm{ord}_{\mathcal{P}}\, K - 1$. The image of ρ in $E/K_{(\mathcal{P})}$ denoted by $(\rho) \in E/K_{(\mathcal{P})}$

<u>is a basis for $\Gamma_{\mathscr{P}}$ as a 1-dimensional $E_{\mathscr{P}}$ vector space.</u>

Now we can begin to do some computations. Assume $\mathscr{P} = \overline{\mathscr{P}}$ is over inert so that the induced involution on $E_{\mathscr{P}}$ is non-trivial. We have $H(E_{\mathscr{P}})$ and a Hermitian form over $E_{\mathscr{P}}$ is denoted by (V,β).

(3.11) <u>Theorem:</u> <u>If $\mathscr{P} = \overline{\mathscr{P}}$ over inert then there is a localizer $\rho_0 \in E^*$ with $\rho_0\overline{\rho}_0^{-1} = u$. Furthermore, there is an isomorphism</u>

$$H(E_{\mathscr{P}}) \simeq H_u(E/K_{(\mathscr{P})})$$

<u>which to (V,β) assigns (V,b) with $b(v,w) = (\beta(v,w)\rho_0) \in E/K_{(\mathscr{P})}$.</u>

<u>Proof:</u>

For any localizer, $\rho\overline{\rho}^{-1} \in O_E(\mathscr{P})^*$. Since we are at an inert $H^1(C_2;O_E(\mathscr{P})^*) = \{1\}$. Hence there is $v \in O_E(\mathscr{P})^*$ with $\rho\overline{\rho}^{-1}v\overline{v}^{-1} = u$. Set $\rho_0 = \rho v$.

Next observe that

$$\overline{ub(v,w)} = \overline{(u\beta(v,w)\rho_0)} = (\overline{u\beta(v,w)\rho_0}) = (\beta(w,v)\rho_0) = b(w,v),$$

as required. This is trivially seen to be a monomorphism. It is an epimorphism by appealing to anisotropic representatives for elements in $H_u(E/K_{(\mathscr{P})})$. ■

According to (II.3.8)

$$\mathrm{rk}\colon H(E_{\mathscr{P}}) \simeq F_2 .$$

Actually we can directly define $\mathrm{rk}\colon H_u(E/K_{(\mathscr{P})}) \to F_2$ for every $\mathscr{P} = \overline{\mathscr{P}}$. The residue field has a certain order q, which is p^f where p is the rational prime under $\mathscr{P}$. If $M_{\mathscr{P}}$ is a finitely generated torsion $O_E(\mathscr{P})$-module then it has order q^r for some $r \geq 0$. We allow the trivial module. Under direct sums orders multiply. If

$(M_{\mathscr{P}}, b)$ has a metabolizer N then $\#(M_{\mathscr{P}})$ is $(\#N)^2$. Thus $rk\colon H_u(E/K_{(\mathscr{P})}) \to F_2$ assigns to $[M_{\mathscr{P}}, b]$ the value

$$\log_q(\#(M_{\mathscr{P}})) \pmod 2.$$

(3.12) <u>Corollary</u>: <u>If</u> $\mathscr{P} = \overline{\mathscr{P}}$ <u>over inert</u>

$$rk\colon H_u(E/K_{(\mathscr{P})}) \simeq F_2 . \blacksquare$$

We turn now to $\mathscr{P} = \overline{\mathscr{P}}$ over ramified. At a ramified prime the induced involution on the residue field $E_{\mathscr{P}}$ is trivial. Therefore $E_{\mathscr{P}} = F_P = O_F(\mathscr{P})/m(P)$ and $H(E_{\mathscr{P}}) = W(F_P)$. We could formally duplicate (III.3.11) if we could find a localizer ρ_0 with $\rho_0\overline{\rho}_0^{-1} = u$. This is not always possible. Recall $H^1(C_2;O_E(\mathscr{P}))^*) \simeq C_2$ generated by $c\ell(\pi\overline{\pi}^{-1})$. Since $\text{ord}_{\mathscr{P}}\,\rho = \text{ord}_{\mathscr{P}}\,K - 1$ we see that for every localizer, ρ, $c\ell(\rho\overline{\rho}^{-1}) = c\ell(\pi\overline{\pi}^{-1})^{\nu_{\mathscr{P}}(K)-1}$.

(3.13) <u>Definition</u>: <u>A finite ramified prime is of type</u> (α) <u>for the pair</u> (u,K) <u>if and only if</u>

$$c\ell(u) = (c\ell(\pi\overline{\pi}^{-1}))^{\nu_{\mathscr{P}}(K)-1}$$

<u>in</u> $H^1(C_2;O_E(\mathscr{P})^*)$.

Certainly this depends both on K and u. We may now state

(3.14) <u>Theorem</u>: <u>At a ramified prime of type</u> (α) <u>there is a localizer</u> ρ_0 <u>with</u> $\rho_0\overline{\rho}_0^{-1} = u$. <u>Furthermore</u>

$$W(F_p) \simeq H_u(E/K_{(\mathscr{P})})$$

by assigning to each symmetric innerproduct space (V,β) over F_p the torsion u-innerproduct

$$b(v,w) = (\beta(v,w)\rho_0) \in E/K_{(\mathscr{P})} \; . \; ■$$

But there must be a second type of ramified prime.

(3.15) Definition: If $\mathscr{P} = \overline{\mathscr{P}}$ is over ramified then $\mathscr{P}$ is of type (β) for (u,K) if and only if $c\ell(u) = c\ell(\pi\overline{\pi}^{-1})^{\nu_{\mathscr{P}}(K)}$ in $H^1(C_2; O_E(\mathscr{P})^*)$.

We can indicate at a non-dyadic ramified the type distinction in terms of the residue field $E_{\mathscr{P}}$. There is a natural isomorphism

$$H^1(C_2; O_E(\mathscr{P})^*) \simeq \{-1,1\} \subset E_{\mathscr{P}}^* \; .$$

If $v \in O_E(\mathscr{P})^*$ is a local unit with $v\overline{v} = 1$ then $v = \pm 1$ in $E_{\mathscr{P}}$. This is the isomorphism.

(3.16) Lemma: A non-dyadic ramified prime is of type (α) for (u,K) if and only if $u = (-1)^{\nu_{\mathscr{P}}(K)-1}$ in $E_{\mathscr{P}}$. Otherwise it is of type (β). ■

Now we can clear up the type (β) case

(3.17) Theorem: If $\mathscr{P} = \overline{\mathscr{P}}$ is over non-dyadic ramified of type (β) then $H_u(E/K_{(\mathscr{P})})$ is trivial. If $\mathscr{P} = \overline{\mathscr{P}}$ is over dyadic ramified of type (β) then rk: $H_u(E/K_{(\mathscr{P})}) \simeq F_2$.

Proof:

At a non-dyadic ramified of type (β) we can surely choose the localizer ρ_0 with $\rho_0\bar{\rho}_0^{-1} = -u$. Then we would identify skew-symmetric forms (V,β) over F_P with torsion u-innerproducts. Since F_P has odd characteristic every skew-inner-product space will contain a metabolizer. ■

Fortunately the dyadic ramified of type (β) will disappear from considerations for another reason.

4. Localizers

For the major results of the last section it should be realized that we made special choices of localizers. This is convenient and quite useful computationally. Yet, as we shall see, it can happen that a localizer is forced on us, thus it is reasonable to study the general situation.

At $\mathscr{P}=\bar{\mathscr{P}}$ the selection of a localizer ρ immediately fixes the choice of basis $(\rho) \in \Gamma_{\mathscr{P}} \subset E/K_{(\mathscr{P})}$ as a vector space over $E_{\mathscr{P}}$. In addition the induced involution on $\Gamma_{\mathscr{P}}$ from that on $E/K_{(\mathscr{P})}$ is

$$\overline{(\lambda\rho)} = (\overline{\lambda\rho})$$

for $\lambda \in O_E(\mathscr{P})$. Thus if $\bar{\rho} = v\rho$ for $v \in O_E(\mathscr{P})^*$ we can write

$$\overline{(\lambda\rho)} = (\bar{\lambda}v\rho).$$

Hence using the identification of $E_{\mathscr{P}}$ with $\Gamma_{\mathscr{P}}$ afforded by (ρ) we find that on $E_{\mathscr{P}}$ we receive the star-involution

$$\lambda^* = \bar{\lambda}v \in E_{\mathscr{P}}.$$

If $\mathscr{P}=\bar{\mathscr{P}}$ over inert then $H^1(C_2;E_{\mathscr{P}}^*) = \{1\}$ so there is a $U \in O_E(\mathscr{P})^*$ with $U\bar{U}^{-1} = v$ in $E_{\mathscr{P}}$. Define the $E_{\mathscr{P}}$-linear map $f\colon E_{\mathscr{P}} \to E_{\mathscr{P}}$ by $f(\lambda) = \lambda U$, then $(f(\lambda))^* = f(\bar{\lambda})$

so that over inert the star involution is equivalent to the involution on $E_{\mathcal{P}}$ induced from conjugation on O_E. Thus we can first identify $H(E_{\mathcal{P}}) \simeq H_1(E/K_{(\mathcal{P})})$ by sending the $E_{\mathcal{P}}$-Hermitian innerproduct space (V, β) to $b(v,w) = (\beta(v,w)U\rho) \in E/K_{(\mathcal{P})}$ and then introducing $H_1(E/K_{(\mathcal{P})}) \simeq H_u(E/K_{(\mathcal{P})})$ by multiplying $b(v,w)$ with a local unit $V \in O_E(\mathcal{P})^*$ with $V\bar{V}^{-1} = u$. This identification over an inert

$$H(E_{\mathcal{P}}) \simeq H_u(E/K_{(\mathcal{P})})$$

does not depend on any of the choices made.

Now at a non-dyadic ramified prime because the involution induced on $E_{\mathcal{P}}$ is the identity, we find the star-involution becomes

$$\lambda^* = \pm \lambda.$$

It depends on whether $v = \rho\bar{\rho}^{-1}$ is +1 or -1 in $E_{\mathcal{P}}$. That is $v = +1$ in $E_{\mathcal{P}}$ if and only if $\mathrm{ord}_{\mathcal{P}}\, \rho = \mathrm{ord}_{\mathcal{P}}\, K - 1$ is even. It is -1 in $E_{\mathcal{P}}$ if and only if $\mathrm{ord}_{\mathcal{P}}\, \rho = \mathrm{ord}_{\mathcal{P}}\, K - 1$ is odd. Correspondingly the involution induced on $\Gamma_{\mathcal{P}}$ is multiplication by ± 1. When the non-dyadic ramified prime is of type (α) then $u = \rho\bar{\rho}^{-1}$ in $E_{\mathcal{P}}$ so $W(F_P) \simeq H_u(E/K_{(\mathcal{P})})$ by sending the symmetric (V, β) to $(\beta(v,w)\rho)$. This isomorphism does however depend on the choice of ρ. We could replace ρ by $U\rho$ where $U \in O_F(P)^*$ and U is not a local norm. This has the effect of changing the isomorphism with multiplication by $[U] \in W(F_P)$.

At non-dyadic ramified of type (β) $u = -\rho\bar{\rho}^{-1}$ in $E_{\mathcal{P}}$ which would force the identification

$$W_S(F_P) \simeq H_u(E/K_{(\mathcal{P})})$$

but $W_S(F_P) = \{0\}$ for a field of odd characteristic.

At dyadic ramified the induced involution on $\Gamma_{\mathcal{P}}$ is the identity so $W(F_P) \simeq H_u(E/K_{(\mathcal{P})}) \simeq F_2$ since $u = 1$ in $E_{\mathcal{P}}$. Obviously this isomorphism does not depend on the choice of localizer.

For an exercise about the classification of ramified primes return to (I.1.4), the definition of ramified primes of types I and II. Show that for $K = O_E$ and $u = -1$ the type (α) ramified will be the type I while the type (β) ramified are the type II. For the pair $(1, O_E)$ all finite ramified primes are of type (β). Replace O_E by $\mathfrak{D}^{-1}_{E/F}$, then the ramified of type (α) for $(1, \mathfrak{D}^{-1}_{E/F})$ are the old type I. By the way, when is $|-1, O_E| = |1, O_E]$ in $\mathrm{Iso}(E/F)$?

5. The inverse different

We are going to pause here to develop an example which has been shown by Stoltzfus [St, sec. 4] to have significant applications in topology. We shall have to first develop some elementary lemmas involving the trace homomorphism of E into Q.

Let V be a finite dimensional vector space over E. We give to $\mathrm{Hom}_Q(V,Q)$ the structure of a vector space over E by

$$(z\psi)(v) \equiv \psi(\bar{z}v)$$

for $z \in E$, $\psi\colon V \to Q$ and $v \in V$. We can use the trace of E over Q

$$\mathrm{tr}_{E/Q}\colon \; E \to Q$$

which is Q-linear to define an E-linear transformation

$$\mathrm{Tr}\colon \mathrm{Hom}_E(V,E) \to \mathrm{Hom}_Q(V,Q)$$

by

$$(\mathrm{Tr}(\varphi))(v) = \mathrm{tr}_{E/Q}(\varphi(v))$$

where $\varphi \in \mathrm{Hom}_E(V,E)$. For $z \in E$

$$(\mathrm{Tr}(z\varphi))(v) = \mathrm{tr}_{E/Q}((z\varphi)(v))$$

$$= \mathrm{tr}_{E/Q}(\varphi(\bar{z}v)) = (z(\mathrm{Tr}(\varphi)))(v).$$

(5.1) Lemma: For a finite dimensional vector space V over E, *Tr:* $\mathrm{Hom}_E(V,E) \simeq \mathrm{Hom}_Q(V,E)$.

Proof:

To say this for $V = E$ is equivalent to the separability of E as an extension of Q. It follows immediately for any finite dimensional vector space over E. ■

Now a similar result involving O_E and Z is not quite what is wanted. Instead we introduce [L, p. 60]

$$\mathcal{D}^{-1} = \{z \mid z \in E,\ \mathrm{tr}_{E/Q}(\lambda z) \in Z,\ \text{all } \lambda \in O_E\}.$$

Then $\mathcal{D}^{-1}$ is a fractional O_E-ideal and is the inverse different of E over Q. Obviously $\overline{\mathcal{D}}^{-1} = \mathcal{D}^{-1}$. We are only concerned with finitely generated O_E-modules and for each such finitely generated O_E-module P there is

$$\mathrm{Tr}\colon \mathrm{Hom}_{O_E}(P,\mathcal{D}^{-1}) \to \mathrm{Hom}_Z(P,Z)$$

which is an O_E-module homomorphism. This may be regarded as a natural transformation between the two contravariant functors. Observe now two O_E-module isomorphisms

$$\mathcal{D}^{-1} \simeq \mathrm{Hom}_{O_E}(O_E, \mathcal{D}^{-1})$$

$$\mathcal{D}^{-1} \simeq \mathrm{Hom}_Z(O_E, Z).$$

The first isomorphism assigns to $z \in \mathfrak{D}^{-1}$ the homomorphism $\varphi_z(\lambda) = \lambda\bar{z}$. The second assigns to z, $tr_{E/Q}(\lambda\bar{z}) = (Tr(\varphi_z))(\lambda)$. We now leave to the reader

(5.2) Lemma: For any finitely generated projective O_E-module

$$Tr\colon \operatorname{Hom}_{O_E}(P, \mathfrak{D}^{-1}) \simeq \operatorname{Hom}_Z(P, Z)$$

as O_E-modules. ■

Really it is not difficult to show the projective assumption can be dropped, but that adds nothing of use to us.

By now it must be clear that there is induced

$$tr_{E/Q}\colon E/\mathfrak{D}^{-1} \to Q/Z$$

and we shall now verify the obvious conjecture.

(5.3) Lemma: If M is a finitely generated O_E-module then

$$Tr\colon \operatorname{Hom}_{O_E}(M, E/\mathfrak{D}^{-1}) \simeq \operatorname{Hom}_Z(M, Q/Z).$$

Proof:

Take a short exact sequence

$$0 \to X \to Y \to M \to 0$$

of finitely generated O_E-modules in which Y and X are projective. Since Y is a finitely generated free abelian group we can, for any additive homomorphism $m\colon M \to Q/Z$, find an additive homomorphism $\psi\colon Y \to Q$ for which the diagram

$$\begin{array}{ccc} Y & \xrightarrow{\psi} & Q \\ \downarrow & & \downarrow \\ M & \xrightarrow{m} & Q/Z \end{array}$$

commutes. In particular $\psi(X) \subset Z$. So by (III.5.2) there is an O_E-module homomorphism $\varphi: X \to \mathfrak{D}^{-1}$ with $Tr(\varphi) = \psi$. But φ extends uniquely to an O_E-module homomorphism $\varphi: Y \to E$. Passing to M and $E/\mathfrak{D}^{-1}$ this will induce an element of $Hom_{O_E}(M, E/\mathfrak{D}^{-1})$ whose image under Tr is $m: M \to Q/Z$.

We must show also that the kernel is trivial. Let $\nu: E \to E/\mathfrak{D}^{-1}$ be the quotient homomorphism. If $\varphi \in Hom_{O_E}(M, E/\mathfrak{D}^{-1})$ and $Tr(\varphi) = 0$ pick any $x \in M$ and a $z \in E$ with $\nu(z) = \varphi(x)$. For any $\lambda \in O_E$ $tr_{E/Q}(\lambda z) \in Z$ because $\nu(\lambda z) = \lambda\nu(z) = \lambda\varphi(x) = \varphi(\lambda x)$. Hence $z \in \mathfrak{D}^{-1}$ and $\varphi(x) = 0 \in E/\mathfrak{D}^{-1}$. ■

Now for the application. Let $\mathcal{P} \subset O_E$ be any finite prime. There is the residue field $E_{\mathcal{P}}$ which is thought of an extension of its prime subfield F_p. Thus we have

$$tr_{E_{\mathcal{P}}/F_p}: E_{\mathcal{P}} \to F_p .$$

Embed $F_p \subset Q/Z$ by sending 1 to $1/p$. We denote the composition by

$$tr_{E_{\mathcal{P}}/F_p}: E_{\mathcal{P}} \to Q/Z.$$

Surely this is an additive homomorphism and $E_{\mathcal{P}}$ is a finitely generated O_E-module. If we now apply (III.5.3) we see

(5.4) <u>Theorem</u>: <u>For each finite prime $\mathcal{P} \subset O_E$ there is a unique O_E-module embedding of $E_{\mathcal{P}}$ into $E/\mathfrak{D}^{-1}$ for which the diagram</u>

$$\begin{array}{ccc} E_{\mathcal{P}} & \longrightarrow & E/\mathfrak{D}^{-1} \\ & \searrow^{tr_{E_{\mathcal{P}}/F_p}} & \downarrow tr_{E/Q} \\ & & Q/Z \end{array}$$

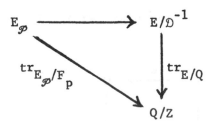

commutes. ■

The image of $E_{\mathscr{P}}$ in $E/\mathfrak{D}^{-1}$ is $\Gamma_{\mathscr{P}}$. We are only concerned with $\mathscr{P}=\overline{\mathscr{P}}$. The image of $1 \in E_{\mathscr{P}}$ is a basis for $\Gamma_{\mathscr{P}}$ and corresponds to a localizer $\rho \in \mathscr{P}^{-1}\mathfrak{D}^{-1}$ for $E/\mathfrak{D}^{-1}$. As usual we write $\nu(\rho) = (\rho) \in E/\mathfrak{D}^{-1}$.

Suppose $\mathscr{P}=\overline{\mathscr{P}}$ is over inert. Then $\overline{\rho} = v\rho$ for some $v \in O_E(\mathscr{P})^*$. On $\Gamma_{\mathscr{P}}$ the induced involution is $(\overline{\lambda\rho}) \to (\overline{\lambda}v\rho)$. But

$$tr_{E/Q}(\lambda\rho) \equiv tr_{E/Q}(\overline{\lambda\rho}) = tr_{E/Q}(\overline{\lambda}v\rho).$$

While in $E_{\mathscr{P}}$ we would conclude that

$$tr_{E_{\mathscr{P}}/F_p}(\lambda-\overline{\lambda}v) = tr_{E_{\mathscr{P}}/F_p}(\lambda-\lambda\overline{v}) \equiv 0.$$

Therefore $1-\overline{v}=0$ in $E_{\mathscr{P}}$ or $v=1$ in $E_{\mathscr{P}}$ and hence $(\overline{\rho}) = (\rho)$ for $\overline{\mathscr{P}}=\mathscr{P}$ over (inert).

We must also comment on the classification of finite ramified primes for the pair $(1,\mathfrak{D}^{-1})$. There is the product formula [L, p. 60], $\mathfrak{D}^{-1} = \mathfrak{D}^{-1}_{E/F}\cdot\mathfrak{D}^{-1}_{F/Q}$ so if $\mathscr{P}=\overline{\mathscr{P}}$ is over ramified

$$\text{ord}_{\mathscr{P}}\, \mathfrak{D}^{-1} \equiv \text{ord}_{\mathscr{P}}\, \mathfrak{D}^{-1}_{E/F} \pmod 2.$$

Thus $\mathscr{P}$ is of type (α) for the pair $(1,\mathfrak{D}^{-1})$ if and only if $\text{ord}_{\mathscr{P}}\, \mathfrak{D}^{-1}_{E/F} \equiv 1 \pmod 2$. That is, if and only if $\mathscr{P}$ is of type I ramified. Clearly the type (β) ramified for $(1,\mathfrak{D}^{-1})$ are the dyadic ramified of type II. Let us close with this lemma.

(5.5) Lemma: If $\mathcal{P} = \overline{\mathcal{P}}$ over non-dyadic ramified then there is a localizer $\rho = \bar{\rho} \in \mathfrak{D}^{-1}\mathcal{P}^{-1}$ for which the resulting embedding of $E_{\mathcal{P}}$ into $E/\mathfrak{D}^{-1}$ produces a commutative diagram

$$\begin{array}{ccc} E_{\mathcal{P}} & \longrightarrow & E/\mathfrak{D}^{-1} \\ & \searrow^{tr_{E_{\mathcal{P}}/F_p}} & \downarrow tr_{E/Q} \\ & & Q/Z\ . \end{array}$$

Proof:

At such a prime $E_{\mathcal{P}} = F_p$ and there is $tr_{F/Q}: F/\mathfrak{D}_{F/Q}^{-1} \to Q/Z$. Now let $N = (p+1)/2$ and there will be a commutative diagram, using (III.5.3)

$$\begin{array}{ccc} E_{\mathcal{P}} & \longrightarrow & F/\mathfrak{D}_{F/Q}^{-1} \\ & \searrow^{N \cdot tr_{E_{\mathcal{P}}/F_p}} & \downarrow tr_{F/Q} \\ & & Q/Z\ . \end{array}$$

This means there is a $\rho \in P^{-1}\mathfrak{D}_{F/Q}^{-1}$ so that for $\lambda \in O_F$

$$tr_{F/Q}(\lambda\rho) = (N \cdot tr_{E_{\mathcal{P}}/F_p}(\lambda))/p \quad \text{in} \quad Q/Z.$$

But $tr_{E/Q}(\lambda\rho) = 2\ tr_{F/Q}(\lambda\rho)$ so the lemma follows. ■

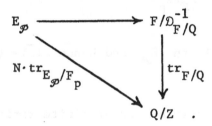

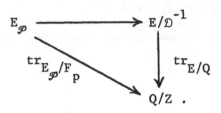

IV The Group $H_u(K)$

1. Basic definitions

We continue with a relative quadratic number field extension E/F. We fix a pair (u,K) wherein $u \in O_E^*$ is a unit with $u\bar{u} = 1$ and $K \subset E$ is a fractional O_E-ideal for which $K = \bar{K}$. We shall consider pairs (P,h) for which

1) P is a finitely generated projective O_E-module

2) $h: P \times P \to K$ is a biadditive function for which

$$\lambda h(v,w) = h(\lambda v,w) = h(v,\bar{\lambda} w)$$

$$\overline{uh(v,w)} = h(w,v)$$

3) The adjoint $Ad_h: P \to Hom_{O_E}(P,K)$ is an isomorphism of O_E-modules.

As usual the O_E-module structure on $Hom_{O_E}(P,K)$ is

$$(\lambda\varphi)(v) = \varphi(\bar{\lambda} v),$$

for all $\lambda \in O_E$, $\varphi \in Hom_{O_E}(P,K)$ and $v \in P$. We should discuss $Hom_{O_E}(P,K)$. The simplest case is $Hom_{O_E}(O_E,K)$. There is a canonical O_E-module isomorphism

$$K \simeq Hom_{O_E}(O_E,K).$$

To $k \in K$ assign $\varphi_k: O_E \to K$ defined by $\varphi_k(\lambda) = \lambda\bar{k}$. What then of the general case? There is the canonical isomorphism

$$Hom_{O_E}(P,O_E) \otimes_{O_E} Hom(O_E,K) \simeq Hom(P \otimes_{O_E} O_E, O_E \otimes_{O_E} K) \simeq Hom_{O_E}(P,K).$$

The isomorphism arises as follows. To $\psi \otimes \varphi$ in the first tensor product it assigns the unique homomorphism

$$\theta\colon P \otimes O_E \to O_E \otimes K$$

for which $\theta(v \otimes \lambda) = \varphi(v) \otimes \varphi(\lambda)$. Obviously we can translate this into

(1.1) <u>Lemma</u>: <u>There is a canonical O_E-module isomorphism</u>

$$\mathrm{Hom}_{O_E}(P,O_E) \otimes_{O_E} K \simeq \mathrm{Hom}_{O_E}(P,K)$$

<u>assigning to</u> $\psi \otimes k$ <u>the unique homomorphism</u> $\theta\colon P \to K$ <u>given by</u> $\theta(v) = \psi(v)\bar{k}$. ■

Suppose that P is taken to be a fractional O_E-ideal $A \subset E$. Then $\bar{A}^{-1} \simeq \mathrm{Hom}_{O_E}(A,O_E)$ so that

$$\bar{A}^{-1}K \simeq \mathrm{Hom}_{O_E}(A,K).$$

In particular, $O_E \simeq \mathrm{Hom}_{O_E}(K,K)$.

Now in general for a finitely generated projective O_E-module there is by Steinitz an isomorphism [O'M, p. 212, 81:5]

$$P \simeq O_E \oplus \ldots \oplus O_E \oplus A$$

where A is a fractional ideal. Accordingly

$$\mathrm{Hom}_{O_E}(P,K) \simeq K \oplus \ldots \oplus K \oplus \bar{A}^{-1}K \simeq O_E \oplus \ldots \oplus O_E \oplus \bar{A}^{-1}K^n.$$

Continuing now with K-valued u-innerproducts (P,h) we say, as usual, for a submodule $L \subset P$ that $L^\perp$ is the submodule of all $v \in P$ with $h(v,L) = u\overline{h(L,v)} = 0$.

(1.2) <u>Lemma</u>: <u>For every submodule</u> $L \subset P$, $L^\perp$ <u>is a direct summand of</u> P.

Proof:

Suppose $v \in P$ and for some λ, $\lambda \neq 0$, $\lambda v \in L^{\perp}$. Then $h(\lambda v, L) = \lambda h(v,L) = 0$. So $\lambda \in L^{\perp}$ and $P/L^{\perp}$ is torsion free and hence 0_E-projective. It follows $L^{\perp}$ is a summand of P. ■ As a consequence we cannot expect $(L^{\perp})^{\perp} = L$ in general, however.

(1.3) Lemma: If L is a summand of P then $L = (L^{\perp})^{\perp}$.

Proof:

This is simply an outline for it is wholly analogous to (III.3.2). If L is a finitely generated projective 0_E-module put $L^* = \mathrm{Hom}_{0_E}(L,K)$ and then $L \simeq L^{**}$ in the usual fashion. If $L \subset P$ is a summand then there is a short exact sequence

$$0 \to L^{\perp} \to P \to L^* \to 0$$

and since K is projective there is a dual short exact sequence

$$0 \to (L^*)^* \to P^* \to (L^{\perp})^* \to 0.$$

Using the adjoint isomorphism this is compared with

$$0 \to (L^{\perp})^{\perp} \to P \to (L^{\perp})^* \to 0$$

and the lemma will follow. ■

We want to introduce Witt equivalence and the group $H_u(K)$. It can be done with as little pain as possible using the following approach. For $V = P \otimes_{0_E} E$, a finite dimensional vector space over E, define

$$h(v \otimes z_1, w \otimes z_2) = h(v,w)z_1\bar{z}_2 .$$

It follows $u\overline{h(v,w)} = h(w,v)$ so (V,h) is a u-innerproduct space over E. We think of $P \subset V$ as $P \otimes 1$.

(1.4) Lemma: The K-valued u-innerproduct (P,h) has a metabolizer if and only if (V,h) does.

Proof:

Let $N \subset P \otimes_{O_E} E$ be a metabolizer for (V,h). Consider then $P \cap N$. For any vector $v \in N$ there is a $\lambda \neq 0$ in O_E with $\lambda v \in P \cap N$. Thus if $v = v \otimes 1 \in P$ lies in $(P \cap N)^{\perp}$ it will follow that $v \in N^{\perp} = N$ and hence $v \in (P \cap N)$. Therefore $P \cap N$ is a metabolizer for (P,h). ■

(1.5) Definition: The two u-innerproduct spaces (P,h) and (P_1,h_1) over K are Witt equivalent if and only if $(P \oplus P_1, h \oplus (-h_1))$ contains a metabolizer.

We need to show this is an equivalence relation, but by (IV.1.4), $(P,h) \sim_W (P_1,h_1)$ if and only if $(P \otimes E,h) \sim_W (P_1 \otimes E,h_1)$. Thus it suffices to consider u-innerproducts over E. We leave this, together with the definitions of $H_u(K)$ and $H_u(E)$ for the reader.

Now we have arrived at groups $H_u(K)$ and $H_u(E)$ together with a natural monomorphism

$$0 \to H_u(K) \to H_u(E).$$

We can relieve any apprehension about $H_u(E)$ by

(1.6) Lemma: The Witt group H(E) is isomorphic to $H_u(E)$.

By Hilbert Theorem 90 there is a $z \in E^*$ with $z = u\bar{z}$. If $(V,\hat{h})$ is an ordinary E-valued Hermitian inner product space, introduce $h(v,w) = z\hat{h}(v,w)$. Then $\overline{uh(v,w)} = \overline{u\bar{z}\,\hat{h}(v,w)} = z\hat{h}(w,v) = h(w,v)$. Conversely a u-innerproduct is converted back to a Hermitian innerproduct by $\hat{h}(v,w) = z^{-1}h(v,w)$. This preserves sums and metabolizers.

A word of caution. This isomorphism does depend on the choice of z. In other words if $z_1 = u\bar{z}_1$ also then $z_1 = zy$, $y \in F^*$ and z, z_1 induce the same isomorphism if and only if $z_1 z^{-1} = y$ is a norm from E^*. ■

2. The group Iso(E/F) again

Clearly we are running around with pairs (u,K) manufacturing Witt groups $H_u(K)$; short of actual computation there is surely some way of deciding when $H_u(K)$ is isomorphic to $H_{u_1}(K_1)$.

(2.1) Theorem: If $|u,K| = |u_1,K_1|$ in Iso(E/F) then there is a canonical family of isomorphisms $H_u(K) \simeq H_{u_1}(K_1)$ in one-to-one correspondence with the elements of the group Gen(E/F).

This is the objective of the present section. We shall use the following lemma.

(2.2) Lemma: Let X, Y, X_1, Y_1 be finitely generated O_E-modules. There is a canonical homomorphism

$$\mathrm{Hom}_{O_E}(X,Y) \otimes_{O_E} \mathrm{Hom}_{O_E}(X_1,Y_1) \to \mathrm{Hom}_{O_E}(X \otimes X_1, Y \otimes Y_1)$$

which is an isomorphism if X and X_1 are projective.

Proof:

The canonical homomorphism is that which to $\varphi \otimes \psi \in \mathrm{Hom}_{O_E}(X,Y) \otimes_{O_E} \mathrm{Hom}_{O_E}(X_1,Y_1)$ assigns the unique homomorphism

$$\theta: X \otimes X_1 \to Y \otimes Y_1$$

for which $\theta(x \otimes x_1) = \varphi(x) \otimes \psi(x_1)$. If $X = X_1 = O_E$ this is obviously an isomorphism. Then it follows immediately for X and X_1 free and finitely generated. By the usual considerations we pass to finitely generated projective. ■

Suppose K and K_1 are involution invariant O_E-fractional ideals and u, u_1 are units with $u\bar{u} = 1 = u_1\bar{u}_1$. Suppose also there is a fractional O_E-ideal $A \subset E$

and an $x \in E^*$ with

$$x A\bar{A} K = K_1$$
$$u x \bar{x}^{-1} = u_1 .$$

Let us then tinker together an isomorphism of $H_u(K)$ with $H_{u_1}(K_1)$. Here is the idea. Begin with a K-valued u-innerproduct (P,h). Form $A \otimes_{0_E} P$ and introduce

$$\hat{h}(a \otimes v, a_1 \otimes v_1) = x a \bar{a}_1 h(v,v_1) \in K_1 .$$

This will define $\hat{h}$ uniquely. Furthermore $u_1 \overline{\hat{h}(a \otimes v, a_1 \otimes v_1)} = u_1 \overline{x a \bar{a}_1 h(v,v_1)}$ $= u x a_1 \bar{a}\, \overline{h(v,v_1)} = x a_1 \bar{a}\, h(v_1,v) = \hat{h}(v_1 \otimes a_1, v \otimes a)$ so $(A \otimes_{0_E} P, \hat{h})$ is now a K_1-valued u_1-innerproduct.

We explain this as follows. Writing $x A\bar{A} = K_1 K^{-1}$ we define (A,β) with

$$\beta(a,a_1) = x a \bar{a}_1 .$$

Since $x\bar{x}^{-1} = u_1 u^{-1}$ this is a $u_1 u^{-1}$-innerproduct with values in $K_1 K^{-1}$. It is an innerproduct because $A = x^{-1}\bar{A}^{-1} K_1 K^{-1} \simeq \mathrm{Hom}_{0_E}(A, K_1 K^{-1})$. Thus we have $[A,\beta] \in H_{u_1 u^{-1}}(K, K^{-1})$. We set up $H_u(K) \to H_{u_1}(K_1)$ by the tensoring with $[A,\beta]$.

In more detail we are invoking these steps. First since A and P are finitely generated projectives we use (IV.2.2) to find

$$\mathrm{Hom}_{0_E}(A, K_1 K^{-1}) \otimes_{0_E} \mathrm{Hom}_{0_E}(P,K) \simeq \mathrm{Hom}_{0_E}(A \otimes_{0_E} P, K_1 K^{-1} \otimes_{0_E} K)$$
$$\simeq \mathrm{Hom}_{0_E}(A \otimes_{0_E} P, K_1).$$

Then the adjoint of (A,β) gives

$$A \simeq \mathrm{Hom}_{0_E}(A, K_1 K^{-1})$$

while the adjoint of (P,h) yields

$$P \simeq \mathrm{Hom}_{O_E}(P,K).$$

Thus we have

$$A \otimes_{O_E} P \simeq \mathrm{Hom}_{O_E}(A \otimes_{O_E} P, K_1)$$

which is precisely what is meant by the adjoint of $(A \otimes_{O_E} P, \hat{h})$. So $\hat{h}$ is a K_1-valued u_1-innerproduct on $A \otimes P$. This process preserves sums and metabolizers so we do receive a homomorphism $H_u(K) \to H_{u_1}(K_1)$.

Yes, it does have an inverse since

$$x^{-1} A^{-1}\bar{A}^{-1} K_1 = K$$

and $u_1 x^{-1}\bar{x} = u$. We had $[A,\beta] \in H_{u_1 u^{-1}}(K_1 K^{-1})$ and so now we have $[A^{-1}, \beta_1]$ in $H_{uu_1^{-1}}(KK_1^{-1})$. Obviously

$$\beta_1(\alpha_1, \alpha_2) = x^{-1}\alpha_1\overline{\alpha_2}$$

for $\alpha_1\alpha_2$ in A^{-1}. If we form a tensor product $(A \otimes_{O_E} A^{-1}, \beta \otimes \beta_1)$ with values in $K_1K^{-1}KK_1^{-1} = O_E$ we receive $[1] \in H_1(O_E)$. The point is

$$(\beta \otimes \beta_1)(a \otimes \alpha, a_1 \otimes \alpha_1) = xa\bar{a}_1\, x^{-1}\alpha\bar{\alpha}_1 = a\bar{a}_1\alpha\bar{\alpha}_1 .$$

But $A \otimes_{O_E} A^{-1} \simeq O_E$ by $a \otimes \alpha \to a\alpha$ so that up to an isometry $(A \otimes A^{-1}, \beta \otimes \beta_1)$ is O_E with the Hermitian innerproduct $(\lambda_1, \lambda_2) \to \lambda_1\bar{\lambda}_2$.

In (IV.2.1) we spoke of a family of isomorphisms between $H_u(K)$ and $H_{u_1}(K_1)$. Let $x_1 B\bar{B}K = K_1$ with $u_1 = ux_1\bar{x}_1^{-1}$. Then $xx_1^{-1} = y \in F^*$ and

$$yAB^{-1}\bar{A}\bar{B}^{-1} = O_E$$

so with $C = AB^{-1}$ we find $\langle y, C\rangle \in \mathrm{Gen}(E/F)$. Now (x,A) and (x_1, A_1) induce the same isomorphism $H_u(K) \simeq H_{u_1}(K_1)$ if and only if y is a norm from E^*, that is, if and only if $\langle y, C\rangle = \langle 1, O_E\rangle$ in $\mathrm{Gen}(E/F)$. ■

Now we are done with (IV.2.1). Agreed that we have not yet computed anything but we can now see the role played by $\mathrm{Iso}(E/F)$ and its relation to $\mathrm{Gen}(E/F)$.

(2.3) Corollary: If for the pair (u,K) the primes $P_1, \ldots, P_r$ are all the finite ramified primes of type (α), then $|1, \mathscr{P}_1 \cdots \mathscr{P}_r| = |u,K|$ and

$$H_u(K) \simeq H_1(\mathscr{P}_1 \cdots \mathscr{P}_r). \quad ■$$

This is a useful device for studying $H_u(K)$. It follows easily by using the isomorphism

$$\mathrm{Iso}(E/F) \simeq \mathscr{R}/\mathscr{R}^2.$$

If at least one prime ramifies, but (u,K) has no finite ramified primes of type (α), then $H_u(K) \simeq H_1(O_E)$. In the unramified case $\mathrm{Iso}(E/F) \simeq Z^*$ so $H_u(K)$ is isomorphic either to $H_1(O_E)$ or $H_1(\mathscr{P})$, where $\mathscr{P} = \overline{\mathscr{P}}$ over inert.

3. The Knebusch exact sequence

Following [M-H, p. 93] we expect to introduce an exact sequence

$$0 \to H_u(K) \to H_u(E) \xrightarrow{\partial} H_u(E/K).$$

The question is the definition of ∂.

(3.1) Exercise: Let (V,h) be a u-innerproduct space over E. Show that there is a finitely generated O_E-module $L \subset V$, containing a basis for V, for which (L,h) is K-valued.

We shall refer to such L as a <u>lattice</u>. For any lattice $L \subset (V,h)$ we define

$$L^{\#} = \{v \mid v \in V, h(L,v) \subset K\}.$$

Again $u\overline{h(L,v)} = h(v,L)$ so that $v \in L^{\#}$ if and only if $h(v,L) \subset L$. There is a natural O_E-module isomorphism

$$L^{\#} \simeq \mathrm{Hom}_{O_E}(L,K)$$

which to $v \in L^{\#}$ assigns

$$\varphi_v(\ell) = h(\ell,v) \in K,$$

for all $\ell \in L$. Of course $L \subset L^{\#}$. We also claim $L \simeq \mathrm{Hom}_{O_E}(L^{\#},K)$. To $\ell \in L$ associate

$$\psi_\ell(v) = u^{-1}h(v,\ell).$$

The explanation is this. We have $L^{\#} \simeq \mathrm{Hom}_{O_E}(L,K)$. Also $L \simeq \mathrm{Hom}_{O_E}(\mathrm{Hom}_{O_E}(L,K),K))$. The effect of $\ell^{**}(\varphi_v)$ is

$$\overline{\varphi_v(\ell)} = \overline{h(\ell,v)} = u^{-1}h(v,\ell) = \psi_\ell(v).$$

On the quotient $L^{\#}/L$ we introduce the biadditive function

$$b\colon L^{\#}/L \times L^{\#}/L \to E/K$$

by

$$b(v+L, w+L) = h(v,w) \pmod{K},$$

where v,w lie in $L^{\#}$.

(3.2) Lemma: <u>This $(L^{\#}/L, b)$ is a torsion u-innerproduct with values in</u> E/K.

Proof:

Denote by $\phi: L^{\#} \to E/K$ and O_E-module homomorphism which vanishes on L. Since $L^{\#}$ is projective and $E \to E/K$ is an O_E-module homomorphism there is an O_E-module homomorphism $\psi: L^{\#} \to E$ for which the diagram

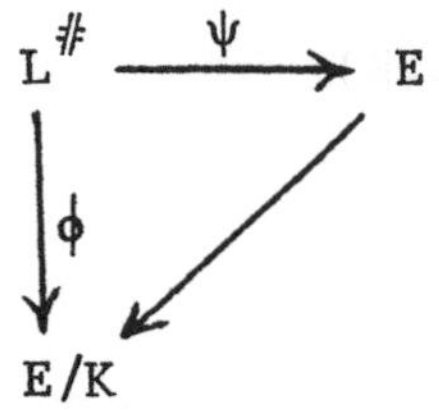

commutes. In particular $\psi(L) \subset K$. Hence there is a $w \in L^{\#}$ with $\psi(\ell) = h(\ell, w)$ for all $\ell \in L$. But then $\psi(v) = h(v,w)$ for all $v \in L^{\#}$. It follows then that

$$Ad_b: L^{\#}/L \to Hom_{O_E}(L^{\#}/L, E/K)$$

is an epimorphism. If there were a kernel element, say $w+L$, $w \in L^{\#}$, then $h(v,w) \in K$ for all $v \in L^{\#}$. Thus we could find $\ell \in L$ with $h(v,w) = u^{-1}h(v,\ell)$ for all $v \in L^{\#}$. Therefore $w = \bar{u}^{-1}\ell$ and since L is an O_E-submodule $w \in L$. ■

(3.3) Lemma: <u>The Witt class of</u> $[L^{\#}/L, b]$ in $H_u(E/K)$ <u>depends only on</u> $[V,h] \in H_u(E)$.

Proof:

Suppose first that V has a metabolizer $N \subset V$. Consider $N \cap L^{\#}$. We claim this is an O_E-module direct summand of $L^{\#}$. If $v \in L^{\#}$ and $\lambda v \in L^{\#} \cap N$ for some $0 \neq \lambda \in O_E$ then $\lambda v \in N^{\perp}$ because $N/N \cap L^{\#}$ is a torsion O_E-module. But then $v \in N^{\perp} = N$ so $v \in N \cap L^{\#}$. So $L^{\#}/N \cap L^{\#}$ is a torsion free O_E-module and hence projective. The assertion follows.

Call $N_1 \subset L^{\#}/L$ the image of $N \cap L^{\#}$ under the quotient homomorphism. We want to show $N_1^{\perp} = N_1$ with respect to b. Obviously $N_1 \subset N_1^{\perp}$. Thus suppose we have a $v \in L^{\#}$ such that $h(N \cap L^{\#}, v) \subset K$. (This is the necessary and sufficient condition that the image of v lie in $N_1^{\perp}$). But v then defines an O_E-module homomorphism of $N \cap L^{\#}$ into K. This is extended to a homomorphism of $L^{\#}$ into K. Hence there is $\ell \in L$ for which

$$u^{-1}h(w,\ell) = h(w,v)$$

for all $w \in N \cap L^{\#}$. Thus $v - \bar{u}^{-1}\ell$ lies in $(N \cap L^{\#})^{\perp} = N \cap L^{\#}$. However, v and $v - \bar{u}^{-1}\ell$ have the same image in $L^{\#}/L$, so the image of v lies in N_1. Thus N_1 is a metabolizer for $(L^{\#}/L, b)$.

Next suppose $(V,h) \sim_W (V_1,h_1)$. Assume lattices $L \subset V$, $L_1 \subset V_1$. Then $L \oplus L_1$ is a lattice for $(V \oplus V_1, h \oplus (-h_1))$ with dual lattice $L^{\#} \oplus L_1^{\#}$ and $(L^{\#}/L \oplus L_1^{\#}/L_1, b \oplus (-b_1))$ results. Since $(V \oplus V_1, h \oplus (-h_1))$ has a metabolizer so does $(L^{\#}/L \oplus L_1^{\#}/L, b \oplus (-b_1))$. ∎

In this fashion we obtain a well defined homomorphism $\partial: H_u(E) \to H_u(E/K)$.

(3.4) Knebusch: The sequence

$$0 \to H_u(K) \to H_u(E) \xrightarrow{\partial} H_u(E/K)$$

is exact.

Proof:

Suppose for some lattice $L \subset (V,h)$ there is a metabolizer $N_1 \subset L^{\#}/L$ with respect to b. Let $T \subset L^{\#}$ be the inverse image of N_1 with respect to the quotient homomorphism. Now $L \subset T \subset L^{\#}$ and the restriction of h to T is K-valued so $T \subset T^{\#}$. Now $L \subset T \subset L^{\#}$ implies $L \subset T \subset T^{\#} \subset L^{\#}$. Since $N_1 = N_1^{\perp}$ and $N_1^{\perp}$ is the image of $T^{\#}$ we find $T = T^{\#}$. Thus T is a "self-dual" lattice and there is $[T,h] \in H_u(K)$ which maps into $[V,h] \in H_u(E)$. ■

(3.5) Exercise: Show that for any lattice L in (V,h) there is a lattice W, $L \subset W \subset V$, for which $(W^{\#}/W,b)$ is anisotropic.

Now we would like to link up the Knebusch exact sequence with the isomorphisms studied in IV.2. Thus assume for pairs (u,K) and (u_1,K_1) there is

$$xA\bar{A}K = K_1$$
$$ux\bar{x}^{-1} = u_1 .$$

We need an isomorphism of $H_u(E/K)$ with $H_{u_1}(E/K_1)$. We shall develop this in two steps. First, there is an O_E-bilinear pairing $K_1K^{-1} \times E \to E$ given by multiplication. This contains $K_1K^{-1} \times K \to K_1$ and so induces $K_1K^{-1} \times (E/K) \to E/K_1$. Then with

$$0 \to K \to E \to E/K \to 0$$

we derive from the bilinear pairings

$$\begin{array}{ccccccccc} 0 \to & K_1K^{-1} \otimes K & \to & K_1K^{-1} \otimes E & \to & K_1K^{-1} \otimes E/K & \to 0 \\ & \downarrow & & \downarrow & & \downarrow & \\ 0 \longrightarrow & K_1 & \longrightarrow & E & \longrightarrow & E/K_1 & \longrightarrow 0 \end{array} .$$

Remember that K_1K^{-1} is O_E-projective! The first two verticle homomorphisms are isomorphisms. Therefore

(3.6) Lemma: There is an O_E-module isomorphism

$$(K_1K^{-1}) \otimes_{O_E} (E/K) \simeq E/K_1 . \quad ■$$

Next, let (M,b) be a torsion u-innerproduct with values in E/K. Form $A \otimes_{O_E} M$ and define $\hat{b}$ with values in E/K_1 by

$$\hat{b}(a \otimes v, a_1 \otimes w) = xa\bar{a}_1 b(v,w).$$

The product $xa\bar{a}_1 \in K_1K^{-1}$ and its product with $b(v,w)$ in E/K is taken by the pairing.

We need to show that the canonical homomorphism

$$\begin{aligned}&\mathrm{Hom}_{O_E}(A,K_1K^{-1}) \otimes_{O_E} \mathrm{Hom}(M,E/K)\\ &\qquad \mathrm{Hom}_{O_E}(A \otimes_{O_E} M, K_1K^{-1} \otimes_{O_E} (E/K))\end{aligned}$$

is an isomorphism. This is not as clear as (IV.2.2) because M is now a torsion module. Thus again we recourse to the short exact sequence

$$0 \to X \to Y \to M \to 0$$

with X and Y both finitely generated and projective. Because E/K is injective

$$0 \to \mathrm{Hom}(M,E/K) \to \mathrm{Hom}(Y,E/K) \to \mathrm{Hom}(X,E/K) \to 0$$

is still exact. Furthermore, $\mathrm{Hom}_{O_E}(A,K_1K^{-1})$ is O_E projective so that

$$\begin{aligned}0 &\to \mathrm{Hom}(A,K_1K^{-1}) \otimes \mathrm{Hom}(M,E/K) \to \mathrm{Hom}(A,K_1K^{-1}) \otimes \mathrm{Hom}(Y,E/K)\\ &\to \mathrm{Hom}(A,K_1K^{-1}) \otimes \mathrm{Hom}(X,E/K) \to 0\end{aligned}$$

remains exact. This last is mapped canonically into the short exact sequence

$$\begin{aligned}0 &\to \mathrm{Hom}(A \otimes M, K_1K^{-1} \otimes E/K) \to \mathrm{Hom}(A \otimes Y, K_1K^{-1} \otimes E/K)\\ &\to \mathrm{Hom}(A \otimes X, K_1K^{-1} \otimes E/K) \to 0.\end{aligned}$$

Now A,X,Y are all finitely generated projectives. Thus finally,

$$\mathrm{Hom}(A,K_1K^{-1}) \otimes \mathrm{Hom}(M,E/K) \simeq \mathrm{Hom}(A \otimes M, (K_1K^{-1}) \otimes E/K).$$

Using the identification $(K_1K^{-1}) \otimes (E/K) \simeq E/K_1$ we finally arrive at

(3.7) Lemma: There is a canonical isomorphism

$$\mathrm{Hom}_{O_E}(A,K_1K^{-1}) \otimes_{O_E} \mathrm{Hom}(M,E/K) \simeq \mathrm{Hom}_{O_E}(A \otimes_{O_E} M, E/K_1),$$

for M a finitely generated torsion O_E-module. ■

Now the adjoint of (A,β) gives $A \simeq \mathrm{Hom}_{O_E}(A,K_1K^{-1})$ while the adjoint of (M,b) gives $M \simeq \mathrm{Hom}_{O_E}(M,E/K)$. Thus $(A \otimes_{O_E} M,\hat{b})$ is a torsion u_1-innerproduct over E/K_1. Just as in section 2 this produces the isomorphism $H_u(E/K) \simeq H_{u_1}(E/K_1)$.

(3.8) Theorem: Associated to

$$x A\bar{A} K = K_1$$

$$u x \bar{x}^{-1} = u_1$$

there is a commutative diagram of isomorphisms

$$\begin{array}{ccccccccc}
0 & \longrightarrow & H_u(K) & \longrightarrow & H_u(E) & \xrightarrow{\partial} & H_u(E/K) \\
 & & \downarrow \simeq & & \downarrow \simeq & & \downarrow \simeq \\
0 & \longrightarrow & H_{u_1}(K_1) & \longrightarrow & H_{u_1}(E) & \xrightarrow{\partial} & H_{u_1}(E/K_1)
\end{array}$$

in which $H_u(E) \simeq H_{u_1}(E)$ is given by $[V,h] \to [V,xh]$.

Proof:

There remains the issue of commutativity with ∂. So in (V,h) over E let $L \subset V$ be a K-lattice. Then $L^{\#} \simeq \mathrm{Hom}_{O_E}(L,K)$. We have the usual isomorphism

$$\mathrm{Hom}_{O_E}(A, K_1K^{-1}) \otimes_{O_E} \mathrm{Hom}_{O_E}(L, K) \simeq \mathrm{Hom}_{O_E}(A \otimes_{O_E} L, K_1).$$

This tells us that if we form $(A \otimes_{O_E} V, \hat{h}) = (\hat{V}, \hat{h})$ then $A \otimes_{O_E} L$ is a K_1-valued lattice for $\hat{h}$ and, using the adjoint of β, the dual lattice is

$$(A \otimes_{O_E} L)^{\#} \simeq A \otimes_{O_E} L^{\#}.$$

Finally, since A is projective,

$$0 \to L \to L^{\#} \to L^{\#}/L \to 0$$

yields

$$0 \to A \otimes_{O_E} L \to A \otimes_{O_E} L^{\#} \to A \otimes_{O_E} (L^{\#}/L) \to 0$$

with $\hat{b}$ as the torsion u_1-innerproduct over E/K_1 representing $\partial[\hat{V}, \hat{h}]$.

For the remark $[\hat{V}, \hat{h}] = [V, xh]$ just apply the construction to a rank one u-innerproduct on E itself. This would be $(z, z_1) \to yz\bar{z}_1$ with $y\bar{y}^{-1} = u$. It is clear $\hat{h}$ on E could only be $(z, z_1) \to xyz\bar{z}_1$. ■

Of course in the sense of (IV.2.1) there is a family of isomorphisms between the Knebusch exact sequences whenever $|u, K| = |u_1, K_1|$ in $\mathrm{Iso}(E/F)$.

4. Localization

We know that

$$H_u(E/K) \simeq \sum_{\mathscr{P}=\bar{\mathscr{P}}} H_u(E/K_{(\mathscr{P})}),$$

so the homomorphism $\partial\colon H_u(E) \to H_u(E/K)$ can be split up into a sum of homomorphisms

$$\partial_{\mathscr{P}}: H_u(E) \to H_u(E/K_{(\mathscr{P})})$$

on for each $\mathscr{P} = \bar{\mathscr{P}}$. This may be interpreted as follows. If $\mathscr{P} = \bar{\mathscr{P}}$ then there is defined $H_u(K_{(\mathscr{P})})$ together with a Knebusch exact sequence

$$0 \to H_u(K_{(\mathscr{P})}) \to H_u(E) \xrightarrow{\partial_{\mathscr{P}}} H_u(E/K_{(\mathscr{P})}).$$

To make the appropriate identification an elementary observation is needed. Namely if $L \subset (V,h)$ is a K-valued lattice then $L_{(\mathscr{P})} = O_E(\mathscr{P})L \simeq O_E(\mathscr{P}) \otimes_{O_E} L$ is a $K_{(\mathscr{P})}$-valued lattice and $L^{\#}_{(\mathscr{P})}$, with respect to $K_{(\mathscr{P})}$ is the localization of $L^{\#}$. Furthermore, $L^{\#}_{(\mathscr{P})}/L_{(\mathscr{P})} = (L^{\#}/L) \otimes_{O_E} O_E(\mathscr{P}) = (L^{\#}/L)_{\mathscr{P}}$. Regarding localization as tensoring with $O_E(\mathscr{P})$ over O_E then this is a standard exercise [B]. ■

Now the idea is to localize at each $\mathscr{P} = \bar{\mathscr{P}}$ and compute the resulting $\partial_{\mathscr{P}}$ in terms of the finite residue fields. This is fine in theory, but to obtain an effective read out of $\partial_{\mathscr{P}}$ for all primes at the same time causes some problems.

The first step is to choose an element $z \in E^*$ with $z\bar{z}^{-1} = u$. This will fix an isomorphism $H(E) \simeq H_u(E)$, for which $[V,h] \to [V,zh]$. If z_1 is a second element with $z_1\bar{z}_1^{-1} = u$ also then the induced isomorphisms of $H(E)$ with $H_u(E)$ agree if and only if $z_1 z^{-1} \in F^*$ is a norm from E^*.

Suppose $\mathscr{P} = \bar{\mathscr{P}}$ over inert. We select a localizer $\rho_0 \in E^*$, (III.3.10), with $\rho_0\bar{\rho}_0^{-1} = u$ and $\text{ord}_{\mathscr{P}}\,\rho_0 = \text{ord}_{\mathscr{P}}\,K - 1$. Then the embedding of the residue field $E_{\mathscr{P}} \to E/K_{(\mathscr{P})}$ by $\lambda \to (\lambda\rho)$ induces $H(E_{\mathscr{P}}) \simeq H_u(E/K_{(\mathscr{P})})$. For $t \in F^*$ we want to compute $\partial_{\mathscr{P}}[zt]$.

Assume a local uniformizer $\pi = \bar{\pi}$ has also been chosen. Write $zt = (z\rho_0^{-1})(\rho_0 t)$, observing $z\rho_0^{-1} \in F^*$ and has an

$$\mathrm{ord}_P\; z\rho_0^{-1} = k.$$

Set $\mathrm{ord}_P\; t = j$ and so write

$$zt = \rho_0\, \pi^{k+j}\, U$$

$U \in O_F(P)^*$. Then

$$[zt] = [\rho_0 U] \qquad k+j \equiv 0 \pmod 2$$

$$[zt] = [\rho_0 \pi U] \qquad k+j \equiv 1 \pmod 2.$$

Since $\mathrm{ord}_{\mathscr{P}}\, \rho_0 = \mathrm{ord}_{\mathscr{P}}\, K - 1$ we can see that $m(\mathscr{P}) \subset O_E(\mathscr{P})$ is a $K_{(\mathscr{P})}$ lattice for the innerproduct $\rho_0\, U x\bar{y}$ on E. Then the dual lattice is $O_E(\mathscr{P})$ with the result that on $O_E(\mathscr{P})/m(\mathscr{P})$ we receive $b(\lambda_1,\lambda_2) = (\rho_0 U \lambda_1 \bar{\lambda}_2)$ in $E/K_{(\mathscr{P})}$, which is anisotropic of rank 1 in F_2. Thus if $k+j \equiv 0 \pmod 2$ then $\partial_{\mathscr{P}}[zt]$ in $H(E_{\mathscr{P}}) \simeq F_2$ is non-trivial. In case $k+j = 1 \pmod 2$ then $O_E(\mathscr{P})$ is the $K_{(\mathscr{P})}$ lattice for $\rho_0 \pi U x\bar{y}$ and it is self-dual. Therefore

$$\partial_{\mathscr{P}}[zt] = 0 \qquad k+j = 1 \pmod 2.$$

Let us next recall

$$(t,\sigma)_P = (-1)^{\nu_{\mathscr{P}}(t)} = (-1)^j$$

$$\mathrm{ord}_{\mathscr{P}}\, z - \mathrm{ord}_{\mathscr{P}}\, \rho_0 = \mathrm{ord}_{\mathscr{P}}\, z - \mathrm{ord}_{\mathscr{P}}\, K + 1 = k.$$

So for $[t]$ in $H(E)$ we can write that

$$\partial_{\mathscr{P}}[zt] = (-1)^{\nu_{\mathscr{P}}(z) + \nu_{\mathscr{P}}(K)}(t,\sigma)_P$$

in $H(E_{\mathscr{P}})$. We are identifying $H(E_{\mathscr{P}})$ with Z^* rather than F_2 so that our formulas will be framed in Hilbert symbols.

In general an element of $H(E)$ may be written $[t_1]+\ldots+[t_r]$. We then find

$$\partial_{\mathscr{P}}([zt_1]+\ldots+[zt_r]) = (-1)^{r(\nu_{\mathscr{P}}(z) + \nu_{\mathscr{P}}(K))}(t_1,\ldots,t_r,\sigma)_P .$$

While $t_1,\ldots,t_r$ would appear as the determinant, we should remember that for an inert prime -1 is a local norm. With that we can state

(4.1) Lemma: If $\mathscr{P}=\overline{\mathscr{P}}$ over inert then for $[V,h] \in H(E)$ the image of $\partial_{\mathscr{P}}([V, zh])$ in Z^* is

$$(-1)^{rk[V,h](\nu_{\mathscr{P}}(z) + \nu_{\mathscr{P}}(K))}(dis[V,h],\sigma)_P . \quad ■$$

In fact this is quite independent of our choice of localizer, but is heavily dependent on both z and K.

Let us next take up $\mathscr{P}=\overline{\mathscr{P}}$ over ramified of type (α), (III.3.12).

(4.2) Lemma: Let $\mathscr{P}=\overline{\mathscr{P}}$ be over ramified of type (α), then there is a localizer ρ_0 for which

1) $\rho_0\bar{\rho}_0^{-1} = u$

2) $z\rho_0^{-1} \in F^*$ is a norm from E^*.

Proof:

Since by definition of type (α) it will follow that for any localizer, ρ,

$$c\ell(\rho\bar{\rho}^{-1}) = c\ell(u) \in H^1(C_2;O_E(\mathscr{P})^*)$$

we can certainly find a localizer with $\rho\bar{\rho}^{-1} = u$. Then $z^{-1}\rho \in F^*$ and since $P \subset O_F$ is ramified we can write $z\rho^{-1} = (\pi\bar{\pi})^k U$ for some $U \in O_F(P)^*$. Thus put $\rho_0 = \rho U$ to obtain the required (localizer).

(4.3) <u>Definition</u>: <u>At a ramified prime of type</u> (α) <u>a localizer that satisfies</u> (IV.4.2) <u>is said to be compatible with</u> z.

We are ready to compute $\partial_{\mathscr{P}}$ using a <u>compatible localizer</u>. With $t \in F^*$ we write $zt = (z\rho_0^{-1})(\rho_0 t)$ so that $[zt] = [\rho_0 t] \in H_u(E)$. Since $t = (\pi\bar{\pi})^j U$ for some $U \in O_F(P)^*$ we need only consider $\partial_{\mathscr{P}}[\rho_0 U]$. Clearly $m(\mathscr{P})$ is the $K_{(\mathscr{P})}$-lattice for $\rho_0 U x\bar{y}$ with dual lattice $O_E(\mathscr{P})$. Then on the residue field $O_E(\mathscr{P})/m(\mathscr{P})$ we receive the torsion u-innerproduct over $E/K_{(\mathscr{P})}$ given by

$$b(\lambda_1,\lambda_2) = (\rho_0 U \lambda_1 \bar{\lambda}_2) \in E/K_{(\mathscr{P})} .$$

Using the embedding of $F_P = E_{\mathscr{P}}$ that sends 1 to (ρ_0) this will identify with $[U] \in W(F_P)$. Now the dyadic case can be completely disposed of by

(4.4) <u>Lemma</u>: <u>At a dyadic ramified prime of type</u> (α)

$$\partial_{\mathscr{P}}[V, zh] \in W(F_P) \simeq F_2$$

is $rk[V,h] \bmod 2$. ■

For the non-dyadic case we think of $W(F_P)$ as $Gp(F_P)$. Furthermore, $Gp(F_P)$ can be regarded as $Z^* \dot{x} F_2$ where product is

$$(d_1,\varepsilon_1)(d_2,\varepsilon_2) = ((-1,\sigma)_P^{\varepsilon_1\varepsilon_2} d_1 d_2 , \varepsilon_1 + \varepsilon_2).$$

Hence [U] in $W(F_P)$ is identified with $((U,\sigma)_P, 1) = ((t,\sigma)_P, 1)$.

Representing an element of [V,h] as a sum $[t_1] + \ldots + [t_r]$ we find

(4.5) Lemma: <u>At a ramified non-dyadic of type (α), using a localizer compatible with z to make the identification,</u>

$$\partial_{\mathscr{P}}[V,zh] = ((\text{dis}[V,h],\sigma)_P , \text{rk}[V,h])$$

<u>in</u> $Gp(F_P)$. ■

This does depend on the choice of localizer. Finally we are left with ramified of type (β). In this case $c\ell(u) \neq c\ell(\pi\bar{\pi}^{-1})^{\nu_{\mathscr{P}}(K)-1}$ so it must follow that $c\ell(u) = c\ell(\pi\bar{\pi}^{-1})^{\nu_{\mathscr{P}}(K)}$. Arguing just as in (IV.4.2) we select a $\theta \in E^*$ such that

1) $\text{ord}_{\mathscr{P}}\, \theta = \text{ord}_{\mathscr{P}}\, K$

2) $\theta\,\bar{\theta}^{-1} = u$

3) $z\theta^{-1}$ is a norm.

Write $zt = z\theta^{-1}\theta(\pi\bar{\pi})^{j}U$ so that $[zt] = [\theta U]$ in $H_u(E)$. But notice $\text{ord}_{\mathscr{P}}\, K = \text{ord}_{\mathscr{P}}\, \theta$ so for $\theta U x\bar{y}$ the $K_{(\mathscr{P})}$-lattice is $O_E(\mathscr{P})$ and it is self-dual.

(4.6) Lemma: <u>At $\mathscr{P} = \bar{\mathscr{P}}$ over ramified of type (β), $\partial_{\mathscr{P}}$ is trivial.</u> ■

Thus the dyadic ramified of type (β) also disappear from our computations. This completes the local computations. To close this section we suggest an exercise.

(4.7) Exercise: If at least one prime ramifies (or in the unramified case if $Sc(c\ell(u)) = 1 \in Z^*$) show there is a z in E^* with $z\bar{z}^{-1} = u$ and $\text{ord}_{\mathscr{P}} z \equiv 0 \pmod 2$ for all $\mathscr{P} = \bar{\mathscr{P}}$ over inert. ■

5. Computing $H_u(K)$

Let us recall from the computation of $Iso(E/F)$ in section I.4 that for any pair (u,K) we can find a pair $(1,R)$ with $|u,K) = |1,R|$ in $Iso(E/F)$. This is convenient because $H_1(R) \to H_1(E) = H(E)$ and we can use $z = 1$.

(5.1) Lemma: If E/F has at least one ramified prime and if $|u,K| = |1,R|$ in $Iso(E/F)$ then the set of finite ramified primes of type (α) for the pair (u,K) agrees with the set of ramified finite primes of type (α) for the pair $|1,R|$. Similarly for type (β). In the unramified case, identifying $Iso(E/F)$ with Z^* by $|1,\mathscr{P}| \to -1$;

$$|u,K| = (Sc(c\ell(u)))(-1)^{\sum \nu_{\mathscr{P}}(K)}$$

with sum over all $\mathscr{P} = \bar{\mathscr{P}}$ over inert. ■

Review the computation of $Iso(E/F)$ in (I.4).

Let us begin with trivialities.

(5.2) Theorem: Assuming no infinite prime ramifies, then the Witt group $H_u(K)$ is trivial in these three cases:

I. E/F is unramified and $Sc(c\ell(u))(-1)^{\sum \nu_{\mathscr{P}}(K)} = -1$.

II. no dyadic prime ramifies, there is at least one ramified prime of type (α) and not more than one ramified of type (β).

III. <u>exactly one dyadic prime ramifies, there is a ramified prime of type</u> (α) <u>and all non-dyadic ramified are of type</u> (α).

<u>Proof</u>:

In case I we have $|u,K| = |1,\mathscr{P}_0|$ where $\mathscr{P}_0 = \overline{\mathscr{P}}_0$ is a chosen prime over an inert P_0. We shall determine $H_1(\mathscr{P}_0) \subset H(E)$ and naturally we take $z = 1$. Looking at (IV.4.1) we shall read

$$\partial_{\mathscr{P}}([V,h]) = (\text{dis}[V,h],\sigma)_P$$

for $\mathscr{P}_0 \neq \mathscr{P} = \overline{\mathscr{P}}$ over inert P, while

$$\partial_{\mathscr{P}_0}[V,h] = (-1)^{\text{rk}[V,h]}(\text{dis}[V,h],\sigma)_{P_0} .$$

So if $\text{rk}[V,h] = 1 \pmod 2$ then $\partial[V,h] \neq 0 \in H_1(E/\mathscr{P}_0)$, for if it were 0 we would have a blatant contradiction to the Hilbert reciprocity theorem. If $\text{rk}[V,h] = 0 \pmod 2$ and $\partial[V,h] = 0$ then from (I.2 2) $\text{dis}[V,h] = 1 \in H^1(C_2;E^*)$ and so $[V,h] = 0$ in $H(E)$.

For part II replace (u,K) by a suitable $(1,R)$, paying attention to (IV.5.1). Appealing to (IV.4.5) we see $\partial[V,h] = 0$ if and only if

1) $\text{rk}[V,h] = 0 \pmod 2$

2) $(\text{dis}[V,h],\sigma)_P = 1$, all P inert or ramified type (α).

Since there is no more than one ramified of type (β) it will follow from reciprocity that $\text{dis}[V,h] = 1$ and thus $[V,h] = 0$.

In part III the dyadic prime may be the only ramified, but then it must be of type (α). Anyway, from $\partial[V,h] = 0$ we get $\text{rk}[V,h] = 0 \pmod 2$. Apply reciprocity to find $\operatorname{dis}[V,h] = 1$. Reciprocity is needed to catch the dyadic ramified no matter of which type. ■

As the reader will have guessed these are the only cases in which the group $H_u(K)$ is trivial. There are several problems with a direct assault on $H_u(K)$. For example, we have found no way to define total signature directly on $H_u(K)$. We could choose z with $z\bar{z}^{-1} = u$ and use the isomorphism $H(E) \simeq H_u(E)$, but this is hardly intrinsic. We can do a little better with discriminant. Of course the rank modulo 2 homomorphism; that is, $\mathrm{rk}\colon H_u(E) \to F_2$ is well defined and hence by restriction so is $\mathrm{rk}\colon H_u(K) \to F_2$. But here is a little hitch.

(5.3) __Lemma:__ __The homomorphism__ $\mathrm{rk}\colon H_u(K) \to F_2$ __is non-trivial if and only if__ $|u,K| = |1,O_E|$ __in__ $\mathrm{Iso}(E/F)$.

__Proof:__

Assume $\mathrm{rk}\colon H_u(K) \to F_2$ is non-trivial. If E/F is unramified we have just ruled out case I of (IV.5.2) and since $\mathrm{Iso}(E/F)$ is cyclic of order 2 in this case we must conclude $|u,K| = |1,O_E|$. If only infinite primes ramify then $\mathrm{Iso}(E/F)$ is trivial.

Thus we come to the case where finite primes do ramify. Using (IV.4.4,4.5) we learn that every finite prime must be of type (β) for the pair (u,K). But then $c\ell(u) = c\ell(\pi\bar{\pi}^{-1})^{\nu_{\wp}(K)} \in H^1(C_2;O_E(\wp)^*)$ at every finite ramified prime. Hence by (I.4.1), $|u,K| = |1,O_E|$.

Now discriminant as a homomorphism into $H^2(C_2;E^*)$ is well defined on the kernel, denoted by J_u, of $\mathrm{rk}\colon H_u(E) \to F_2$. In other words put

$$\operatorname{dis}[V,zh] = \operatorname{dis}[V,h]$$

if $\dim V \equiv 0 \pmod 2$. Should we replace z by a z_1 with $z_1\bar{z}_1^{-1} = u$ also then $z_1 = zt$ for some $t \in F^*$. However

$$[V, z_1h] = [V, zth]$$

and since rank is even we find

$$\mathrm{dis}[V, th] = t^{2k}\mathrm{dis}[V, h] = \mathrm{dis}[V, h]$$

in $H^2(C_2; E^*)$.

(5.4) <u>Theorem: The image of the homomorphism</u>

$$\mathrm{dis}: J_u \cap H_u(K) \to H^2(C_2; E^*)$$

<u>consists of those $c\ell(y)$ for which</u>

$$(y,\sigma)_P = 1, \quad \text{all } P \text{ inert}$$
$$(y,\sigma)_P = 1, \quad \text{all } P \text{ non-dyadic ramified type } (\alpha).$$

<u>Proof:</u>

As usual replace (u,K) with $(1.R)$ and take $z = 1$. By (IV.4.1,4.5) we see the image of $\mathrm{dis}: J_u \cap H_u(K) \to H^2(C_2; E^*)$ is contained in the indicated subgroup. If $y \in F^*$ is an element satisfying the requisite conditions then $([y]-[1])$ lies in $J \subset H(E)$ with discriminant $c\ell(y)$. Furthermore $\partial([y]-[1]) = 0$ so $[y]-[1]$ lies in $J \cap H_1(R)$. ■

(5.5) Corollary: The image of

$$\text{dis}: J \cap H_1(O_E) \to H^2(C_2;E^*)$$

agrees with the image of the embedding

$$1 \to \text{Gen}(E/F) \to H^2(C_2;E^*)$$

which is given by $\langle y,A\rangle \to c\ell(y) \in H^2(C_2;E^*)$. ■

In fact the image of

$$\text{dis}: J_u \cap H_u(K) \to H^2(C_2;E^*)$$

is always a subgroup of the image of Gen(E/F).

(5.6) Theorem: If for the pair (u,K) the integer $n \geq 0$ denotes the total number of finite dyadic ramified primes in E/F together with the number of non-dyadic ramified of type (β) for (u,K) then $J_u \cap H_u(K)$ is torsion free if and only if $n \leq 1$. If $n > 1$ then the torsion subgroup of $J_u \cap H_u(K)$ is an elementary abelian 2-group of order 2^{n-1}. Unless $|u,K| = |1,O_E|$ and E/F has no signatures this completely describes the torsion subgroup of $H_u(K)$.

Proof:

Again we work with (1,R). A torsion element in $J \subset H(E)$ is completely determined by its discriminant. Suppose $[V,h] \in J$ is a torsion element. If $\partial[V,h] = 0$ then we know $(\text{dis}[V,h],\sigma)_P = 1$ for all P inert, non-dyadic finite type (α) ramified, and infinite ramified. The only thing we can do then to get a torsion class in $J \cap H_1(R)$ is to choose $y \in F^*$ with $(y,\sigma)_P = 1$ at the above listed primes and $(y,\sigma)_P$ arbitrarily specified, subject to Hilbert reciprocity!, at dyadic ramified and non-dyadic ramified of type (β). Then $\partial([y]-[1]) = 0$ so $[y]-[1]$ gives a torsion class in $J \cap H_1(R)$. So the count is right on the torsion subgroup's order. If $|u,K| \neq |1,O_E]$ then $J_u \cap H_u(K) = H_u(K)$. If signatures are present then the torsion subgroup of $H(E)$ lies in J. ■

(5.7) Addendum: *If no signatures are present then* $H_1(O_E) \simeq F_2$ *if* $n \leq 1$. *If* $n > 1$ *the order of* $H_1(O_E)$ *is* 2^n. *The group contains an element of order* 4; *namely,* [1] *if and only if* -1 *fails to be a norm from* E^*. ■

We have really not said much about the free part of $H_u(K)$. Actually the result is quite simple.

(5.8) Theorem: *If* E/F *has signatures then for any pair* (1,R)

$$\text{Sgn}: J^2 \cap H_1(R) \simeq 4Z(E).$$

Furthermore

$$\text{Sgn}: J \cap H_1(R) \to 2Z(E)$$

is an epimorphism unless every finite ramified prime is non-dyadic and of type (α). *for* (1,R) *in which case the cokernel is cyclic of order* 2

Proof:

First, if $[V,h] \in J^2$ then $rk[V,h] = 0 \pmod 2$ and $1 = dis([V,h])$ in $H^2(C_2;E^*)$. Surely it follows $\partial[V,h] = 0$ and so J^2 lies in the image of $H_1(R) \to H(E)$.

Now go back to (4.9 of II). If there is a dyadic ramified or a non-dyadic ramified of type (β) select one of these and call if P_0. To each real ramified infinite prime P_∞ attach an even integer. Now claim there is $X \in J$ with

$$(\text{dis } X, \sigma)_P = 1$$

for all finite P inert or non-dyadic ramified of type (α) for the pair $(1,R)$ and with $\text{sgn}_{P_\infty} X = 2n_{P_\infty}$. The point is we can set the Hilbert symbol at P_0 as we wish to satisfy reciprocity. Well then $\partial X = 0$ and so X lies in the image of $H_1(R) \to H(E)$.

In the exceptional case we would need $(\text{dis } X, \sigma)_P = 1$ at all finite primes. Hence to realize the assignment of signatures we would need the condition $n_{P_{\infty_1}} + \ldots + n_{P_{\infty_r}} \equiv 0 \pmod 2$ by (II.4.9). This accounts for the cokernel. ■

A few examples might be suggested. Begin with cyclotomic number fields $Q(\xi_n)$ with complex conjugation. The fixed field is $Q(\xi_n + \xi_n^{-1})$ which is completely real and all real infite primes ramify in $Q(\xi_n)/Q(\xi_n + \xi_n^{-1})$. If n is odd composite no finite prime ramifies. If $n = p^k$, p a prime, then exactly one finite prime ramifies. If p is odd then of course it is of type (α) for $(-1, 0_E)$. Neither $H_{-1}(0_E)$ or $H_1(0_E)$ has torsion.

For quadratic extensions of Q many effects are possible. For example if $m > 0$ is square free and $m \equiv 1 \pmod 4$ then $H_{-1}(0_E) = \{0\}$ while $H_1(0_E) \simeq F_2$. A more specific case is $Q(\sqrt{3})$ which has class number 1 so $H^1(C_2; 0_E^*) \simeq \text{Iso}(E/F) \simeq C_2 \oplus C_2$. There is $(1, 0_E)$, $(-1, 0_E)$, $(2 + \sqrt{3}, 0_E)$ and $(-2 - \sqrt{3}, 0_E)$. For $(-1, 0_E)$ the prime 3 is ramified of type (α) while 2 is ramified of type (β), so $H_{-1}(0_E) = 0$. Let $\mathcal{P}_2, \mathcal{P}_3$ denote the primes over 2 and 3. We know $|-1, 0_E| = |1, \mathcal{P}_3|$. Which of the two fundamental units gives $|1,\mathcal{P}_2|$? The other gives $|1,\mathcal{P}_2\mathcal{P}_3|$. The hint is this. You are looking for the fundamental unit with $c\ell(u) \neq c\ell(\pi\bar{\pi}^{-1})^{1-1} = 1 \in H^1(C_2;0_E(\mathcal{P}_3)^*)$. You want $c\ell(u) = c\ell(-1)$ so which fundamental unit is -1 in the residue field $E_{\mathcal{P}_3}$?

Another good quadratic example is $Q(\sqrt{-1})$. Work out $H_u(0_E)$ for $u = \pm 1, \pm i$. The only finite ramified prime is 2 so $\mathrm{Iso}(E/F) \simeq C_2$, generated by $|i,0_E|$.

In (I.5) we pointed out some unramified examples. We noted cases with $Sc(-1) = -1$ and cases with Sc trivial. In the unramified case $H_1(\mathcal{P}) = 0$ for any $\mathcal{P} = \bar{\mathcal{P}}$ over inert and $H_1(0_E) \simeq F_2$.

6. The ring $H(0_E)$

Using tensor products $H_1(0_E) = H(0_E)$ becomes a commutative ring with identity; a subring of $H(E)$. We begin out study with a (diagonalization) result. Recall from (I.3) that the group $\mathrm{Gen}(E/F)$ arose from pairs (y,A) with $yA\bar{A} = 0_E$, $y \in F^*$ and A a fractional 0_E-ideal. This means A supports an 0_E-module Hermitian innerproduct; namely, $(a, a_1) \to ya\bar{a}_1$. Surely this is an innerproduct because $y^{-1}\bar{A}^{-1} = A$ and $y^{-1}\bar{A}^{-1} \simeq \bar{A}^{-1} \simeq \mathrm{Hom}_{0_E}(A,0_E)$. Thus by analogy with the embedding of $H^2(C_2;E^*) \to H(E)^*$ there is an embedding

$$1 \to \mathrm{Gen}(E/F) \to H(0_E)^*$$

which to $\langle y,A\rangle$ assigns the Witt class of the Hermitian innerproduct on A. We shall denote this by $[y,A] \in H(0_E)$, however remember that $\langle y,A\rangle$, and hence $[y,A]$, is uniquely determined by $c\ell(y) \in H^2(C_2;E^*)$, (I.3.3).

(6.1) <u>Theorem: Except for the case in which only two real infinite primes ramify in E/F, the image of Gen(E/F) in $H(0_E)^*$ generates $H(0_E)$ as an additive group.</u>

Proof:

If $X \in H(O_E)$ and $rk(X) = 1 \in F_2$ then $X - [1,O_E] \in J \cap H(O_E)$. Turn then to $X \in J \cap H(O_E)$. For such an X we have $dis(X) = cl(y) \in H^2(C_2;E^*)$ and $(y,\sigma)_P = 1$ at all inert primes. (For the pair $(1,O_E)$ all finite ramified primes are obviously of type (β).) However by (I.3.1) there is then a fractional O_E-ideal A with $yA\bar{A} = O_E$. Thus $([y,A]-[1,O_E])$, certainly in the additive subgroup generated by the image of $Gen(E/F)$, and X have the same discriminant in $H^2(C_2;E^*)$. If no signatures are present we are done. Next let signatures be present. There are two cases.

Case 1: A finite ramified prime occurs in E/F.

In this case we can find elements $y \in F^*$ with arbitrarily prescribed signs at the real infinite primes and with $(y,\sigma)_P = 1 = (-1)^{\nu_P(y)}$ for all inert primes. Therefore the image of $Sgn\colon H(O_E) \to Z(E)$ will agree with the image of the restriction of Sgn to the additive subgroup generated by $Gen(E/F)$.

Case 2: No finite prime ramifies.

In this case an even number of real infinite primes ramify. We must assume that this number is at least four. Number off these ramified real infinite primes: $P_{\infty_1}, \ldots, P_{\infty_{2m}}$. Fix an index i. We note that we can also select indices j and k so that i,j,k are distinct. Now there is $y \in F^*$ with y negative at P_{∞_i}, P_{∞_j} and positive at the ramining orderings and $\nu_P(y) \equiv 0 \pmod 2$ at all inert primes. There is $y_1 \in F^*$ with similar properties for $P_{\infty_i}, P_{\infty_k}$. Now look at the product

$$Y_i = ([y,A]-[1,O_E])([y_1,A_1]-[1,O_E]) .$$

This lies in the subgroup generated by the image of $Gen(E/F)$ and $Sgn_{P_i} Y_i = 4$

while $Sgn_{P_j} Y_i = 0$ for $i \neq j$. Thus the image of $Sgn: J \cap H(O_E) \to Z(E)$ will coincide with the restriction of Sgn to the elements of even rank in the additive subgroup generated by the image of Gen(E/F).

In the exceptional case of exactly two real infinite primes, and no finite primes, ramified the group $H(O_E)$ is free abelian of rank 2 while the image of Gen(E/F) generates an infinite cyclic group. ■

We can form $Gp(O_E)$ by introducing on $Gen(E/F) \times F_2$ the product

$$(\langle y,A\rangle,\varepsilon)(\langle y_1,A_1\rangle,\varepsilon_1) = (\langle(-1)^{\varepsilon\varepsilon_1} yy_1, AA_1\rangle, \varepsilon+\varepsilon_1) .$$

We can also describe

$$H(O_E) \to G_p(O_E) \to 1$$

directly. If (P,h) is an O_E-valued Hermitian innerproduct space then there is the associated innerproduct space over E, $(P \otimes_{O_E} E, h)$. Here $h(v \otimes z, v_1 \otimes z_1) = h(v,v_1)z\bar{z}_1$. The rank of (P,h) is $\dim_E (P \otimes_{O_E} E)$, and so $rk: H(O_E) \to F_2$ is directly defined.

Write $(V,h) = (P \otimes E, h)$. If $n = \dim V$ and $e_1, \ldots, e_n$ is a basis for V then we receive, using $e_1 \wedge \ldots \wedge e_n$, a specific isomorphism $\wedge^n V \simeq E$. This identifies the induced innerproduct structure on $\wedge^n V$ with an Hermitian innerproduct on E which must have the form $yz\bar{z}_1$ for a unique $y \in F^*$. Of course $cl((-1)^{n(n-1)/2}y)$ in $H^2(C_2;E^*)$ is the discriminant of (V,h). Now $P \subset V$ as $P \otimes 1$ and (P,h) also induces an O_E-valued Hermitian innerproduct on $\wedge^n P \subset \wedge^n V$. The isomorphism $\wedge^n V \simeq E$ then identifies $\wedge^n P$ with some fractional O_E-ideal $A \subset E$ for which $yA\bar{A} = O_E$. Hence $[P,h]$ goes to

$$(\langle(-1)^{n(n-1)/2}y, A\rangle, rk[P,h])$$

in $Gp(O_E)$. This is well defined because $cl(y) = 1 \in H^2(C_2;E^*)$ if and only if

$\langle y,A\rangle = \langle 1,O_E\rangle \in Gen(E/F)$. We shall denote by $J_0 \subset H(O_E)$ the kernel of rk: $H(O_E) \to F_2$. Using (IV.6.2) we can argue just as in the field case (II.3) that

(6.2) Corollary: Outside of the exceptional case there is a short exact sequence

$$0 \to J_0^2 \to H(O_E) \to G_p(O_E) \to 1 .$$

There is an isomorphism

$$dis: J_0/J_0^2 \simeq Gen(E/F) .$$

If no signatures are present

$$H(O_E) \simeq Gp(O_E) .$$

If signatures are present

$$Sgn: J_0^2 \simeq 4Z(E) . \blacksquare$$

This is invalid in the exceptional case. In that case it is easy to see that a basis for $H(O_E)$ is given by $[1,O_E]$ and a Witt class G which has $rk(G) = 0 \in F_2$ and $dis\ G = 1 \in Gen(E/F)$ while

$$Sgn_{P_{\infty_1}} G = 4$$

$$Sgn_{P_{\infty_2}} G = 0 .$$

Thus $G^2 = 4G$. Since any element in J_0 has the form $2m[1,O_E] + nG$ it is seen that G does not lie in J_0^2. One can represent G by a rank 4 innerproduct space over O_E.

For the simplest examples of the exceptional case let m_1, m_2 be square free negative relatively prime integers with $m_1 \equiv 1 \pmod 4$. Set $F = Q(\sqrt{m_1 m_2})$ and $E = F(\sqrt{m_1}) = F(\sqrt{m_2})$. The two real infinite primes in F ramify since $m_1 < 0$ but no finite prime can ramify (I.5).

We can sharpen (IV.6.1) with the following observation.

(6.3) Hamrick: Suppose that in E/F at least one finite prime is ramified and that -1 is not a norm from E. Then $H(O_E)$ is additively generated by $N-1$ elements from Gen(E/F) where N is the total number of ramified primes.

Proof:

Because -1 is not a norm we find by reciprocity that $N \geq 2$. Look first at Gen(E/F). A minimal generating set for this group contains $N-1$ elements. If no more than one signature is present choose a minimal generating set S which contains the element $\langle -1, O_E \rangle$. If more than one real infinite prime ramifies number these $P_{\infty_1}, \ldots, P_{\infty_r}$. Since at least one finite prime ramifies there are elements $\langle y_1, A_1 \rangle, \ldots, \langle y_{r-1}, A_{r-1} \rangle$ such that $y_j < 0$ at P_{∞_j} and positive for all other orderings. Adjoin $\langle -1, O_E \rangle$ to this set and complete it to form a minimal generating set S of Gen(E/F).

Let $\tilde{S} \subset H(O_E)$ denote the additive subgroup of the Witt ring generated by the image of S in $H(O_E)^*$. The objective is to show that $\tilde{S}$ contains the image of Gen(E/F) and apply (IV.6.1).

Let $\langle y, A \rangle$, $\langle y', A' \rangle$ be two elements of Gen(E/F) for which both $[y, A]$ and $[y', A']$ do belong to $\tilde{S}$. Compare $[yy', AA']$ with the sum $([y,A] + [y',A'] + [-1, O_E])$. This latter belongs to $\tilde{S}$ since $[-1, O_E] \in \tilde{S}$. Thus the two elements of $H(O_E)$ have rank 1 (mod 2) and the same discriminant $\mathrm{cl}(yy') \in H^2(C_2; E^*)$. Therefore $[yy', AA']$ differs from $([y,A] + [y',A'] + [-1, O_E])$ by an element of J_0^2. If E/F is without signatures then $J_0^2 = \{0\}$ and we are done for then we have shown that if $[y,A]$, $[y',A']$ both belong to $\tilde{S}$ then so does their product and S was a minimal generating set for $Gen(E/F)$.

If signatures are present we make the following observations. If exactly one signature is present then $Sgn\, 2[-1,O_E] = -2$ and therefore $Sgn|\tilde{S} \cap J_0^2 \simeq 4Z$ so that $\tilde{S} \supset J_0^2$ and we are again done. If more than one signature is present then for $1 \leq j \leq r-1$ the $Sgn[y_j,A_j] + [-1,O_E])$ is the vector $-2e_j$ where e_j is the standard base element of Z^r with 1 in the j-th co-ordinate and 0 otherwise. Now $Sgn(2[-1,O_E]) = -2(\sum_{j=1}^{r} e_j)$. Thus $-2e_1,\ldots,-2e_r$ are all in the image of $Sgn|\tilde{S} \cap J_0$. Hence $\tilde{S} \supset J_0^2$ and we are finally done. ■

Our algebraic picture of $H(O_E)$ is now quite analogous to that of $H(E)$. But there is one major difference. From (I.3) we recall the homomorphism j_2: $Gen(E/F) \to H^1(C_2;C(E))$. In fact this extends to a homomorphism $Gp(O_E) \to H^1(C_2;C(E))$ by $(\langle y,A\rangle,\varepsilon) \to c\ell(|A|) \in H^1(C_2;C(E))$. So finally we must arrive at a homomorphism

$$H(O_E) \to H^1(C_2;C(E)).$$

This may be given the following interpretation.

(6.4) **Theorem:** *For* $X \in H(O_E)$ *the following are equivalent:*

(i) X *lies in the kernel of*

$$H(O_E) \to H^1(C_2;C(E))$$

(ii) <u>there is an O_E-Hermitian innerproduct space</u> (P,h), <u>in which</u> P <u>is a finitely generated free O_E-module, with</u> $[P,h]=X$

(iii) <u>there is a unit</u> $u \in O_F^*$ <u>for which</u> $c\ell(u) = \text{dis } X \in H^2(C_2;E^*)$.

<u>Proof</u>:

First suppose P is a finitely generated projective O_E-module, with $V = P \otimes_{O_E} E$. An invariant of P in $C(E)$ arises by choosing a basis $e_1, \ldots, e_n$ and identifying $\Lambda^n P$ with a fractional O_E-ideal $A \subset C(E)$. It is $P \to |A| \in C(E)$ which yields the isomorphism of the reduced projective class group of O_E with $C(E)$; that is,

$$\tilde{K}_0(O_E) \simeq C(E).$$

The Steinitz structure theorem for finitely generated projective O_E-modules shows that P is free if and only if $\Lambda^n P$ is free, if and only if $A \subset E$ is principal.

Now at this point it is clear that (ii) implies (i) for if we have (P,h) representing X with P an O_E-free module we simply choose an O_E-module basis $e_1, \ldots, e_n$ for P to discover $\Lambda^n P \simeq O_E$. This also shows (ii) implies (iii).

Suppose in general we have selected a basis $e_1, \ldots, e_n$ for $(V,h) = (P \otimes_{O_E} E, h)$ and we find $\Lambda^n P$ identifies with (y,A) where $y \in F^*$ and $yA\bar{A} = O_E$. Suppose further that $c\ell(|A|) = 1$ in $H^1(C_2;C(E))$.

Then for some $z \in E^*$ and fractional O_E-ideal B we have

$$z^{-1}B\bar{B}^{-1} = A,$$

and $y = z\bar{z}u$, $u \in O_F^*$. The trick now is this. Since $\bar{B} \simeq \text{Hom}_{O_E}(B^{-1}, O_E)$ we can define on

$$B^{-1} \oplus \overline{B}$$

a Hermitian O_E-valued innerproduct by

$$\hat{h}((b_1,\beta_1),(b_2,\beta_2)) = b_1\overline{\beta}_2 + b_2\overline{\beta}_1 ,$$

where $b_1, b_2 \in B^{-1}$ and β_1, β_2 in $\overline{B}$. This obviously has as metabolizer $(B^{-1},0)$.

Therefore $[P,h] = [P \oplus (B^{-1} \oplus \overline{B}), h \oplus \hat{h}] = X \in H(O_E)$. It is our claim that $P \oplus (B^{-1} \oplus \overline{B})$ is O_E-free. First $B^{-1} \oplus \overline{B} \simeq O_E \oplus B^{-1}\overline{B} \simeq O_E \oplus A^{-1}$. This last follows from $z^{-1}B\overline{B}^{-1} = A$ or $A^{-1} = zB^{-1}\overline{B}$. But the Steinitz structure theorem says

$$P \simeq (\text{Free}) \oplus A.$$

Finally $((\text{Free}) \oplus A) \oplus O_E \oplus A^{-1} \simeq (\text{Free}) \oplus O_E \oplus O_E$.

Thus we find (i) if and only if (ii). To see (i) if and only if (iii) we appeal to the exact hexagon (I.4.4). ■

A quite elementary example will illustrate an extreme situation. Suppose $m > 0$ is a square free integer and that every odd prime p dividing m is congruent to 1 (mod 4). Using (I.2.4) it is an easy exercise to see that $-1 \in Q = F$ is a norm from $E = Q(\sqrt{m})$. (Use reciprocity if 2 divides m.) Therefore $H^2(C_2;O_E^*) \to \text{Gen}(E/F)$ is trivial and $j_2\colon \text{Gen}(E/F) \to H^1(C_2;C(E))$ is a monomorphism. Actually $j_2\colon \text{Gen}(E/F) \simeq H^1(C_2;C(E))$ by application of (I.6.1) to this situation. In any case we find in these examples that if (P,h) is a Hermitian innerproduct space of even rank over O_E for which P is O_E-free then $[P,h] = 0 \in H(O_E)$.

Finally let us consider the relation of all the $H_u(K)$ to the ring $H(O_E)$. Using tensor products, $H_u(K)$, $H_u(E)$ and $H_u(E/K)$ all become $H(O_E)$-modules. Furthermore, the Knebusch sequence

$$0 \to H_u(K) \to H_u(E) \xrightarrow{\partial} H_u(E/K)$$

is easily seen to be an exact sequence of $H(O_E)$ modules and homomorphisms. This suggests that, assuming at least one prime in F ramifies, we assign to each pair (u,K) an ideal in $H(O_E)$ as follows.

1. If for (u,K) there are no finite primes ramified of type (α) put

$$I(u,K) = H(O_E).$$

2. If there are finite primes ramified of type (α) for (u,K) but these are all dyadic put

$$I(u,K) = J_0 \subset H(O_E).$$

3. If $P_1, \ldots, P_s$ are all the finite ramified non-dyadic of type (α) for (u,K) put

$$I(u,K) = \{X \mid X \in J_0 (\text{dis } X, \sigma)_{P_j} = 1,\ 1 \le j \le s\}.$$

It may not be immediately obvious that $I(u,K)$ is an ideal in the third case, but we shall see that it is. Anyway we can agree that $I(u,K)$ only depends on the equivalence class $|u,K| \in \text{Iso}(E/F)$ and $I(u,K) \supset J_0^2$.

We are to see that as $H(O_E)$ modules, $H_u(K) \simeq I(u,K)$. We do not claim this is canonical. We take a pair $(1,R)$ with $|1,R| = |u,K|$, and we assume $R \subset O_E$ is an integral ideal which is the product of all $\mathcal{P} = \overline{\mathcal{P}}$ over P ramified of type (α)

for (u,K). For each such $\mathscr{P}$, $\text{ord}_{\mathscr{P}} R = 1$. We do think of $(1,R)$ as a canonical representative of $|u,K|$. This allows $R = O_E$ if there are no ramified of type (α). Furthermore to study $H_1(R)$ we can take $z = 1$ and for each type (α) ramified we can use $(1) \in E/R_{(\mathscr{P})}$ for compatible localizer. Thus we have

$$0 \to H(O_E) \longrightarrow H(E) \xrightarrow{\partial} H(E/O_E)$$
$$0 \to H_1(R) \longrightarrow H(E) \xrightarrow{\partial_R} H(E/R).$$

We do claim that $\ker(\partial_R) \subset \ker(\partial)$ as $H(O_E)$-modules because ∂ and ∂_R are both $H(O_E)$-module homomorphisms and $\ker(\partial)$ consists of all $X \in H(E)$ with $(\text{dis } X, \sigma)_P = 1$ for all $P \subset O_F$ inert, while $\ker(\partial_R)$, presuming $R \neq O_E$, consists of all $X \in H(E)$ with

$$\text{rk } X = 0 \in F_2$$
$$(\text{dis } X, \sigma)_P = 1, \quad \text{all } P \text{ inert}$$
$$(\text{dis } X, \sigma)_P = 1, \quad \text{all } P|R \text{ non-dyadic.}$$

Using

$$\begin{array}{ccc} H(O_E) & \simeq & \ker(\partial) \\ & & \uparrow \text{inclusion} \\ H_1(R) & \to & \ker(\partial_R) \end{array}$$

we embed $H_1(R)$ into $H(O_E)$ with image the deal $I(u,K) = I(1,R)$. It is not accurate to say $H_1(R) \subset H(O_E)$ for we do not wish to suggest that a Hermitian innerproduct space over R is simultaneously a Hermitian innerproduct space over O_E.

This does give us a handle on one question. It is also clear that there is an $H(O_E)$ bilinear pairing of $H_u(K)$ with $H_{u_1}(K_1)$ into $H_{uu_1}(KK_1)$; that is a homomorphism

$$H_u(K) \otimes_{H(O_E)} H_{u_1}(K_1) \to H_{uu_1}(KK_1).$$

Actually this corresponds to the product of ideals; that is, to $I(u,K)\cdot I(u_1,K_1) \subset I(uu_1,KK_1)$. The containment is correct. Think of it in terms of $(1,R)$ and $(1,R_1)$. The canonical representative of $|1,RR_1|$ will drop out all primes dividing RR_1 with multiplicity 2. One thing is obvious. If $I(u,K)$ and $I(u_1,K_1)$ are both proper ideals then $I(u,K)\cdot I(u_1,K_1) \subset J_0^2$ which has no torsion. In particular, if E/F has no signatures then $I(u,K)\cdot I(u_1,K_1) = \{0\}$. We can translate (IV.5.8) into this formalism as follows.

(6.5) Lemma: Suppose E/F has signatures, and that $I(u,K) \subset J_0$, then the restriction

$$\text{Sgn}: I(u,K) \to 2Z(E)$$

is an epimorphism unless all the finite ramified primes are non-dyadic and all of these are of type (α) for (u,K) in which case the image of

$$\text{Sgn}: I(u,K) \to 2Z(E)$$

consists of those $(2n_1, \ldots, 2n_r)$ with $n_1 + \ldots + n_r \equiv 0 \pmod 2$. ■

Thus our most frequently expected product for proper ideals is $I(u,K)\cdot I(u_1,K_1) = J_0^2$. Show that if $I(u,K) \subset J_0$ then $I(u,K)$ is a prime ideal if and only if $J_0 = I(u,K)$, [M-H, p. 66].

7. Cokernel of ∂

Let us begin by introducing the reduced group $\tilde{H}_u(E/K)$. We shall agree

$$\tilde{H}_u(E/K) = \sum_{P \text{ inert}} H_u(E/K_{(\mathscr{P})}) \oplus \sum_{\substack{P \\ \text{ramified of} \\ \text{type } (\alpha)}} H_u(E/K_{(\mathscr{P})}).$$

Recall $\partial_{\mathscr{P}}$ is trivial at $\mathscr{P} = \overline{\mathscr{P}}$ over ramified of type (β) so it is pointless to consider such primes. Corresponding to this definition there will be $\delta\colon H_u(E) \to \tilde{H}_u(E/K)$ which is just the $\sum \partial_{\mathscr{P}}$ over the allowed primes.

(7.1) Lemma: Suppose for (u,K) that one or more of the following hypotheses are satisfied:

(i) E/F has signatures

(ii) a dyadic prime ramifies

(iii) a non-dyadic prime of type (β) ramifies. Then, if there is no more than one type (α) ramified, δ is an epimorphism but if $m > 1$ is the number of type (α) ramified then coker (δ) is an elementary abelian 2-group of order 2^{m-1}.

Proof:

We choose z with $z\bar{z}^{-1} = u$ and assume compatible localizers for all ramified of type (α). In this fashion we receive the identification

$$\tilde{H}_u(E/K) \simeq \sum_{\text{inert}} H(E_{\mathcal{P}}) \oplus \sum_{\substack{\text{ramified} \\ \text{type } (\alpha)}} W(F_P)$$

and this contains

$$\sum_{\text{inert}} H(E_{\mathcal{P}}) \oplus \sum_{\substack{\text{non-dyadic} \\ \text{type } (\alpha)}} J_P .$$

Here $J_P \subset W(F_P)$ is the fundamental ideal. Remember J_P is trivial for dyadic ramified of type (α).

We first assert

$$\delta: J_u \to \sum H(E_{\mathcal{P}}) \oplus \sum J_P$$

is an epimorphism. Under any one of the three hypotheses we can find an element $y \in F^*$ with the Hilbert symbols arbitrarily prescribed at the inert primes and the non-dyadic ramified of type (α). But then with $X = [zy] - [z]$ in J_u we have

$$\partial_{\mathcal{P}} = (y, \sigma)_P$$

at all primes under consideration (IV.4.1,4.5). We have identified J_P with Z^* as usual.

We now only concern ourselves with $\sum W(F_P)$, since rank mod 2 is preserved at all type (α) ramified (including dyadics). Again we find $y \in F^*$ so that its Hilbert symbols are

$$(y, \sigma)_P = (-1)^{\nu_{\mathcal{P}}(z) + \nu_{\mathcal{P}}(K)}$$

for all P inert and $(y,\sigma)_P$ arbitrarily prescribed at non-dyadic ramified of type (α). By (4.1,4.4,4.5 of IV)

$$\partial_{\mathcal{P}}([y]) = 0 \in H(E_{\mathcal{P}}),\ \mathcal{P}=\bar{\mathcal{P}} \text{ over inert}$$

$$\partial_{\mathcal{P}}([y]) \neq 0 \in W(F_P),\ P \text{ ramified dyadic of type } (\alpha)$$

$$\partial_{\mathcal{P}}([y]) = ((y,\sigma)_P,1) \in W(F_P),\ P \text{ non-dyadic ramified of type } (\alpha).$$

The lemma now follows. ■

Let us take up the unramified case, for the result is a bit surprising.

(7.2) <u>Lemma</u>: <u>If</u> E/F <u>is unramified the coker</u> ($\eth$) <u>is cyclic of order</u> 2 <u>if</u> $|u,K| = |1,O_E| \in \mathrm{Iso}(E/F)$, <u>while</u> $\eth: H_u(E) \simeq \tilde{H}_u(E/K)$ <u>if</u> $|u,K| \neq |1,O_E|$.

<u>Proof</u>:

Let us begin by considering the effect of $\eth$ on J_u. For $X \in J_u$ we know $\eth_{\mathcal{P}}(X) = (\mathrm{dis}\ X,\sigma)_P$ at all inert primes. Thus, by reciprocity and realization, we find that $\eth: J_u \to \tilde{H}_u(E/K)$ has a cokernel which is cyclic of order 2. On the other hand for $y \in F^*$

$$\eth_{\mathcal{P}}[zy] = (-1)^{\nu_{\mathcal{P}}(z) + \nu_{\mathcal{P}}(K)} (y,\sigma)_P$$

for all P inert. Hence

$$\prod_{\substack{P \\ \text{inert}}} (\eth_{\mathcal{P}}[zy]) = (-1)^{\sum \nu_{\mathcal{P}}(z) + \nu_{\mathcal{P}}(K)}.$$

We know $(-1)^{\sum \nu_{\mathscr{P}}(z) + \nu_{\mathscr{P}}(K)} = 1 \in Z^*$ if and only if $|u,K| = |1,0_E|$ in Iso(E/F). In that case $\partial: H_u(E) \to \tilde{H}_u(E/F)$ and $\partial: J_u \to \tilde{H}(E/K)$ have exactly the same image dictated by the reciprocity law. However, if $(-1)^{\sum \nu_{\mathscr{P}}(z) + \nu_{\mathscr{P}}(K)} = -1$ we catch everything in the image of ∂. Also using (IV.5.2) we can conclude $\partial: H_u(E) \simeq \tilde{H}(E/K)$ if $|u,K| \neq |1,0_E|$ in the unramified case. ■

Finally then E/F has no signatures and there are finite primes that ramify but they are non-dyadic of type (α) for (u,K). Because we eventually construct a rather (bizarre) group, we shall give a sketch first. Start again with $\partial: J_u \to \sum H(E_{\mathscr{P}}) \oplus \sum J_P$. Because $\partial_{\mathscr{P}} X = (\text{dis } X, \sigma)_P$ at all P inert or ramified we must conclude that

$$\partial: J_u \to \sum H(E_{\mathscr{P}}) \oplus \sum J_P$$

has a cokernel that is cyclic of order 2. This time ∂ will preserve rank mod 2 at all the ramified primes, so this cokernel will not disappear. Thus let G be the quotient of $\tilde{H}_u(E/K)$ by the image of

$$\partial | J_u \to \tilde{H}_u(E/K).$$

The image of

$$\sum H_u(E/K_{(\mathscr{P})}) \oplus \sum J_P$$

in G will be cyclic of order 2. Furthermore, the image of the composite homomorphism

$$H_u(E) \xrightarrow{\delta} \tilde{H}_u(E/K) \to G$$

is also cyclic of order 2 and is generated by the image of $[z]$ in $H_u(E)$. So we find

(7.3) <u>Lemma</u>: <u>If</u> E/F <u>has no signatures and all finite ramified primes are non-dyadic of type</u> (α) <u>for</u> (u,K) <u>then the coker</u> (δ) <u>is an abelian group of order</u> 2^m <u>where</u> $m \geq 1$ <u>is the number of finite ramified primes. The cokernel contains an element of order</u> 4 <u>if and only if</u> -1 <u>fails to be a norm from</u> E^*. ■

We can be more explicit about the group G. Before we start we comment that it follows from reciprocity that there must be an even number of finite ramified primes for which -1 is not a square in the residue field. With this in mind number off the ramified primes $P_1, \ldots, P_{2r}$, $P'_1, \ldots, P'_s$ so that -1 is a non-square in F_{P_j}, $1 \leq j \leq 2r$ and -1 is a square in $F_{P'_j}$, $1 \leq j \leq s$. We have $r = 0$ if and only if -1 is a norm from E^*. We could have $s = 0$, but then $r > 0$.

Let

$$G = Z^* \times (F_2)^m$$

with product

$$(c; a_1, \ldots, a_{2r}, b_1, \ldots, b_s) \cdot (c'; a'_1, \ldots, a'_{2r}, b'_1, \ldots, b'_s)$$
$$= ((-1)^{a_1 a'_1 + \ldots + a_{2r} a'_{2r}} cc'; a_1 + a'_1, \ldots, a_{2r} + a'_{2r}, b_1 + b'_1, \ldots, b_s + b'_s).$$

(We said it would look strange!) Now using compatible localizers $H_u(E/K_{(\mathcal{P})}) \simeq W(F_P)$ for P ramified and $W(F_P)$ was identified with $Z^* \times F_2$ given the product

$$(d_1\varepsilon_1)\cdot(d_2,\varepsilon_2) = ((-1,\sigma)_P^{\varepsilon_1\varepsilon_2} d_1 d_2, \varepsilon_1\varepsilon_2).$$

Obviously, following the enumeration, we have a natural embedding of $W(F_P)$ into G for each ramified prime. For P inert embed $H(E/K_{(\mathcal{P})})$ into G by sending the non-trivial element to $(-1;0,\ldots,0)$. This produces an epimorphism

$$\tilde{H}_u(E/K) \to G \to 1.$$

Furthermore, using (IV.4.1,4.5) the kernel of this epimorphism is seen to be the image of

$$\delta|J_u \to \tilde{H}_u(E/K).$$

In G, $\delta[z]$ becomes

$$((-1)^{\sum \nu_{\mathcal{P}}(z) + \nu_{\mathcal{P}}(K)}; 1,\ldots,1)$$

where the sum reaches over all P inert. This is clear because

$$\delta_{\mathcal{P}}[z] = (-1)^{\nu_{\mathcal{P}}(z) + \nu_{\mathcal{P}}(K)}(1,\sigma)_P$$

for all P inert while

$$\partial_{\mathcal{P}}[z] = ((1,\sigma)_P, 1) = (1,1) \in G_p(F_P)$$

at every ramified. The element $\delta[z] \in G$ generates a cyclic group of order 2 since $\delta[z] \to (\pm 1;1,\ldots,1)$. The coker (δ) now is the quotient of G by the subgroup that $\delta[z]$ generates.

It is interesting to note that coker (δ) is not related to ideal class groups in contrast to [M-H, p. 94]. For example with $F = Q$, $E = Q(\sqrt{21})$ the coker (δ) associated with the pair $(-1, O_E)$ is cyclic of order 4 although $Q(\sqrt{21})$ is a Euclidean field.

V The Witt Ring $W(O_F)$

1. Symbols

A complete discussion of the Witt ring of symmetric innerproduct spaces over the ring of algebraic integers is to be found in [M-H]. We reproduce some of the results here to allow the reader an opportunity for comparison and contrast with the Hermitian case.

Let F be an algebraic number field and $W(F)$ the Witt ring of symmetric innerproduct spaces over F. As before there are the invariants

$$rk: W(F) \to Z/2Z$$

and

$$dis: W(F) \to F^*/F^{**} .$$

Indeed we may define $Gp(G)$ as

$$F_2 \times F^*/F^{**}$$

with addition

$$(\varepsilon_1, x) + (\varepsilon_2, y) = (\varepsilon_1 + \varepsilon_2 , (-1)^{\varepsilon_1 \varepsilon_2} xy)$$

and

$$(\varepsilon_1, x) \cdot (\varepsilon_2, y) = (\varepsilon_1 \varepsilon_2 , x^{\varepsilon_2} y^{\varepsilon_1})$$

so that by introducing

$$X \to (rk(X), dis(X))$$

there will result a multiplicative short exact sequence

$$0 \to J^2 \to W(F) \to Gp(G) \to 0 .$$

As always $J \subset W(F)$, the fundamental ideal, consists of the Witt classes with even rank. Although we had not previously mentioned a product in $Gp(F)$ it is not difficult to show $W(F) \to Gp(F)$ is multiplicative.

To facilitate the discussion of signatures we remind the reader of

(1.1) Definition: A field is formally non-real if and only if -1 can be expressed as a sum of squares. Otherwise the field is formally real.

Formally non-real fields are characterized by their lack of orderings. Thus an algebraic number field is formally non-real if and only if all its infinite primes are complex. In a formally non-real field the minimal number of squares required to express -1 as a sum is called the Stufe or level. Among algebraic number fields this Stufe may have values 1, 2 or 4.

Assume that F is formally real. Then to each real infinite prime P_∞ there is associated a signature homomorphism

$$\mathrm{Sgn}_{P_\infty} : W(F) \to Z .$$

By forming $Z(F)$, a direct sum of Z with itself, one copy for each real infinite prime, there will result the total signature

$$\mathrm{Sgn}: W(F) \to Z(F) .$$

Up to this point nothing new has been introduced into consideration. Now the Hasse-Minkowski invariant appears to create the difference between this and the Hermitian case. Denote by M_F the set of all primes, finite and infinite, in F. Let S be the multiplicative group of all functions

$$f: M_F \to Z^*$$

such that

1) $f(P) = 1$ for almost all primes

2) $f(P) = 1$ for all complex infinite primes

3) $\prod_P f(P) = 1$.

For elements x,y in F^*/F^{**} we denote by $((x,y)) \in S$ the function which to each prime assigns the value of the Hilbert symbol of x and y at that prime. This yields a symmetric bimultiplicative pairing of F^*/F^{**} with itself into S. The realization theorem for Hilbert symbols states that for any $f \in S$ there is a pair x,y in F^*/F^{**} for which

$$((x,y)) = f \ .$$

Suppose now that (V,b) is an innerproduct over F. If $\dim V = 1$ we put $c(V,b) = 1$. If $\dim V > 1$ choose an orthogonal basis $e_1, \ldots, e_n$ with $\alpha_j = b(e_j,e_j)$ and define $c(V,b) = \prod_{i<j} ((\alpha_i,\alpha_j))$ in S. This yields a well defined invariant of the isometry class of (V,b) and is called the Hasse-Minkowski invariant. The following should be compared with (II.4.4).

(1.2) Hasse: The isometry class of the symmetric innerproduct space (V,b) over F is uniquely determined by

1) rank (V,b)

2) $\operatorname{dis}(V,b)$ in F^*/F^{**}

3) Hasse-Minkowski invariant

4) total signature if F is formally real. ■

We shall summarize the well known basic properties of $c(V,b)$ in

(1.3) Lemma: For two innerproduct spaces (V,b), (V_1,b_1) over F

1) $c(V \oplus V_1, b \oplus b_1) = ((\det(V,b), \det(V_1,b_1)))c(V,b)c(V_1,b_1)$.

2) <u>if</u> $\dim V \equiv \dim V_1 \equiv 0 \pmod 4$

$$c(V \otimes V_1, b \otimes b_1) = ((\det(V,b), \det(V_1, b_1)))$$

3) <u>if</u> $\dim V = 2k$ and (V,b) <u>is metabolic then</u> $c(V,b) = ((-1,-1))^{k(k-1)/2}$. ■

We must put this into the form of a Witt class invariant. Given $X \in W(F)$ select a representative innerproduct space (V,b) with $\dim V \equiv 0$ or $1 \pmod 8$. Let $c(X) = c(V,b) \in S$.

(1.4) <u>Lemma</u>: <u>The invariant</u> $c\colon W(F) \to S$ <u>is well defined.</u>

<u>Proof</u>:

Assume (V_1, b_1) is a second such representative of X with $\dim V_1 \equiv \dim V \pmod 8$. First, from our choices of dimension and the fact that (V,b), (V_1,b_1) are Witt equivalent it follows that $\det(V,b) = \operatorname{dis}(V,b) = \operatorname{dis}(V_1,b_1) = \det(V_1,b_1)$. Next, introduce $(H_1,\beta_1) = (V_1 \oplus V_1, -b_1 \oplus b_1)$ and $(H_0,\beta_0) = (V \oplus V_1, b \oplus -b_1)$. Then $(V \oplus H_1, b \oplus \beta_1)$ is isometrically equivalent to $(H_0 \oplus V_1, \beta_0 \oplus b_1)$ and hence these have the same Hasse-Minkowski invariant. Since (H_0,β_0) and (H_1,β_1) are both metabolic with $\dim H_0 \equiv \dim H_1 \pmod 8$ we can say

$$\det(H_0,\beta_0) = \det(H_1,\beta_1)$$

$$c(H_0,\beta_0) = c(H_1,\beta_1) .$$

It now follows that $c(V,b) = c(V_1,b_1)$. ■

Next there is the task of relating this invariant to addition and multiplication in $W(F)$.

(1.5) <u>Lemma</u>: <u>Let</u> X <u>and</u> Y <u>be elements of</u> $W(F)$ <u>with</u> $\operatorname{rk}(X) = \varepsilon_1 \in Z/2Z$, $\operatorname{rk}(Y) = \varepsilon_2$ in $Z/2Z$. <u>Then</u>

$c(X+Y) = ((-\text{dis } X \text{ dis } Y, (-1)^{\varepsilon_1\varepsilon_2}))((\text{dis } X, \text{dis } Y))\, c(X)c(Y)$ <u>and</u> $c(X\cdot Y) = ((\text{dis } X, \text{dis } Y))^{\varepsilon_1\varepsilon_2+1}\, c(X)^{\varepsilon_2} c(Y)^{\varepsilon_1}$.

Proof:

Choose representatives (V,b) and (V_1,b_1) with appropriate dimensions (mod 8). Note that $\det(V,b) = \text{dis } X$, $\det(V_1,b_1) = \text{dis } Y$. Thus

$$c(V \oplus V_1, b \oplus b_1) = ((\text{dis } X, \text{dis } Y))c(X)c(Y).$$

Now if $\varepsilon_1\varepsilon_2 = 0$ in $Z/2Z$ then $V \oplus V_1$ has the correct dimension (mod 8) to represent $X+Y$ and we are done. If $\varepsilon_1\varepsilon_2 = 1$ we must add a six dimensional metabolic which will have invariant $((-1,-1))$ and determinent -1. Now $\det(V \oplus V_1, b \oplus b_1) = \text{dis } X \text{ dis } Y$. Thus $c(X+Y) = ((-1,-1))((-1, \text{dis } X \text{ dis } Y))c(V \oplus V_1, b \oplus b_1)$. Applying the bimultiplicative property of Hilbert symbols the formula for addition follows.

Now suppose X and Y both lie in J. We may suppose then that $\dim V \equiv \dim V_1 \equiv 0 \pmod 8$. Thus by (V.3.5)

$$c(XY) = c(V \otimes V_1, b \otimes b_j) = ((\text{dis } X, \text{dis } Y)).$$

Now suppose $Y = \langle 1 \rangle + W$ with $W \in J$. Then

$$XY = X + XW$$

so that

$$c(XY) = ((\text{dis } X, \text{dis } XW))c(X)c(XW).$$

However $XW \in J^2$ so $\text{dis } XW = 1$. Now then

$$c(XY) = c(X)((\text{dis } X, \text{dis } W)).$$

Again $W \in J$ so $\operatorname{dis} Y = \operatorname{dis} W$.

Finally suppose $X = \langle 1 \rangle + U$, $Y = \langle 1 \rangle + W$, U, W in J. Then

$$XY = X + XW$$

and with $XW \in J$

$$c(XY) = ((\operatorname{dis} X, \operatorname{dis} XW))\, c(X)c(XW) .$$

Furthermore

$$\operatorname{dis}(XW) = \operatorname{dis} W = \operatorname{dis} Y$$

while

$$c(XW) = ((\operatorname{dis} X, \operatorname{dis} W))c(X)c(W) .$$

Finally,

$$c(W) = c(Y)$$

so

$$c(XY) = c(X)c(Y) .$$

The formula given is seen to cover all the cases. ■

We should translate (V.1.2) into the determination of $W(F)$.

(1.6) Theorem: An element of $W(F)$ is uniquely determined by

1) rank mod 2

2) discriminent

3) Hasse-Minkowski invariant

4) total signature if F is formally real. ■

We knew that dis is trivial on J^2 and using (V.1.5) c is trivial on J^3 and so induces a homomorphism $c\colon J^2/J^3 \to S$. This is an isomorphism. First for $f \in S$ there is a pair x,y in F^*/F^{**} with $((x,y)) = f$. Let $X = \langle x\rangle - \langle 1\rangle$, $Y = \langle y\rangle - \langle 1\rangle$ so that

$$c(XY) = ((\text{dis } X, \text{dis } Y)) = ((x,y)) .$$

If F is formally non-real then $J^3 = \{0\}$ simply because we ran out of invariants. If F is formally real then

$$\text{Sgn}\colon J^3 \to 8Z(F)$$

must be a monomorphism. In fact it is an epimorphism. Fix a real infinite prime P_∞. There is a pair x,y in F^*/F^{**} with $((x,y))(P_\infty) = 1$ and $((x,y)) = -1$ at all other real infinite primes. Thus $x < 0$, $y < 0$ for all real infinite primes escept P_∞. Say $x > 0$ at P_∞ and consider $-x$. Thus $\text{Sgn}_{P_\infty}(\langle -x\rangle - \langle 1\rangle) = -2$ and all other signatures are 0. Using $(\langle -x\rangle - \langle 1\rangle)^3$ we can see that

$$\text{Sgn}\colon J^3 \to 8Z(F) .$$

We can now complete the argument that

$$c\colon J^2/J^3 \simeq S .$$

We shall need the lemma which relates total signature to the other invariants (II.4.8).

(1.7) Lemma: Suppose F is formally real then for every real infinite prime

$$Sgn_{P_\infty} X \equiv rk(X) \bmod 2,\ X \in W(F);\quad ((\mathrm{dis}\ X_1 - 1))(P_\infty) = (-1)^{(Sgn_{P_\infty} X)/2},\quad X \in J$$

$$c(X)(P_\infty) = (-1)^{(Sgn_{P_\infty} X)/4},\quad X \in J^2 .$$

<u>Proof</u>:

The first case is clear. If $X \in J$ then $X = (\langle \mathrm{dis}\ X\rangle - \langle 1\rangle) + Y$ with $Y \in J^2$ so $\mathrm{dis}\ Y = 1$, $Sgn_{P_\infty} Y \equiv 0 \pmod 4$. Thus $Sgn_{P_\infty}(X) \equiv 2 \pmod 4$ if and only if $\mathrm{dis}\ X < 0$ at P_∞. For $X \in J^2$ we can write, using realization,

$$X = (\langle x\rangle - \langle 1\rangle)(\langle y\rangle - \langle 1\rangle) + Y$$

with $Y \in J^3$ so that $c(Y) = 1$, $c(X) = ((x,y))$. Now $Sgn_{P_\infty}(Y) \equiv 0 \pmod 8$ so that

$$Sgn_{P_\infty}(X) = (Sgn_{P_\infty}(\langle x\rangle - \langle 1\rangle))(Sgn_{P_\infty}(\langle y\rangle - \langle 1\rangle)) \equiv 4 \pmod 8$$

if and only if $((x,y))(P_\infty) = -1$. ■

Now if $X \in J^2$ and $c(X) = 1$ it follows $Sgn(X) \in 8Z(F)$ and for some $Y \in J^3$, $Sgn(X-Y) = 0$ and $c(X-Y) = 1$ of course so $X = Y \in J^3$.

We can introduce a ring $Sym(F)$ which enlarges $Gp(G)$. On

$$F_2 \times F^*/F^{**} \times S$$

we introduce

$$(\varepsilon_1, x, f_1) + (\varepsilon_2, y, f_2) = (\varepsilon_1 + \varepsilon_2, (-1)^{\varepsilon_1\varepsilon_2} xy, ((-xy, (-1)^{\varepsilon_1\varepsilon_2}))((x,y)) f_1 f_2)$$

and

$$(e_1, x, f_1)\cdot(\varepsilon_2, y, f_2) = (\varepsilon_1\varepsilon_2, x^{\varepsilon_2} y^{\varepsilon_1}, ((x,y))^{\varepsilon_1\varepsilon_2 + 1} f_1^{\varepsilon_2} f_2^{\varepsilon_1}) .$$

(1.8) <u>Theorem</u>: <u>For any algebraic number field the ring homomorphism</u>

$$X \to (\mathrm{rk}(X), \mathrm{dis}(X), c(X))$$

yields a short exact sequence

$$0 \to J^3 \to W(F) \to \mathrm{Sym}(F) \to 0 .$$

If F is formally non-real

$$W(F) \simeq \mathrm{Sym}(F) .$$

If F is formally real

$$\mathrm{Sgn}: J^3 \simeq 8Z(F) . \blacksquare$$

We should observe that

$$\langle x \rangle \to (1,x,1)$$
$$\langle x \rangle - \langle 1 \rangle \to (1,x,1) + (1,-1,1) = (0,x,1)$$
$$(\langle x \rangle - \langle 1 \rangle)(\langle y \rangle - \langle 1 \rangle) \to (0,1,((x,y))) .$$

If F is formally non-real W(F) is a torsion group. If F is formally real then the torsion ideal in W(F) is the kernel of

$$\mathrm{Sgn}: W(F) \to Z(F)$$

and so the torsion ideal embeds isomorphically into Sym(F).

(1.9) Lemma: If F is formally real then the torsion ideal in W(F) is isomorphic to the ideal of elements (0,x,f) where $x \in F^*/F^{**}$ is positive with respect to every ordering and $f(P_\infty) = 1$ for all real infinite primes.

Proof:

First suppose that X is a torsion element so that since $\mathrm{Sgn}(X) = 0$ it will follow that $\mathrm{rk}(X) = 0$. Write $X = \langle \mathrm{dis}\ X \rangle - \langle 1 \rangle + Y$ with $Y \in J^2$. Then for each real infinite prime $0 = \mathrm{Sgn}_{P_\infty}(\langle \mathrm{dis}\ X\rangle - \langle 1\rangle) + \mathrm{Sgn}_{P_\infty}(Y)$. However, $\mathrm{Sgn}_{P_\infty}(Y) \equiv 0 \pmod 4$ while $\mathrm{Sgn}_{P_\infty}(\langle \mathrm{dis}\ X\rangle - \langle 1\rangle) = 0$ or -2. Thus we concluded

$$\mathrm{Sgn}(\langle \mathrm{dis}\ X\rangle - \langle 1\rangle) = 0$$
$$\mathrm{Sgn}(Y) = 0\ .$$

It must follow that $\mathrm{dis}(X) > 0$ for every ordering and from (IV.1.7) $c(Y)(P_\infty) = 1$ for all real infinite primes. According to our addition formula

$$c(X) = c(\langle \mathrm{dis}\ X\rangle - \langle 1\rangle)c(Y)\ .$$

But $c(\langle \mathrm{dis}\ X\rangle - \langle 1\rangle)(P_\infty) = 1$ at all real infinite primes by (V.1.7). Thus $c(X)(P_\infty) = 1$.

Now suppose $x \in F^+/F^{**}$ and $f \in S$ has $f(P_\infty) = 1$ at all real infinite primes. Choose $Y \in J^2$ with $c(Y) = f$. Then there is $Y_1 \in J^3$ with $\mathrm{Sgn}(Y) = \mathrm{Sgn}(Y_1)$ so that $Y - Y_1$ is a torsion class. Next $\langle x\rangle - \langle 1\rangle$ is seen to have signature 0 since x is totally positive and thus $\langle x\rangle - \langle 1\rangle + Y - Y_1$ is a torsion class mapping into $(0, x, f)$.

If $S_T \subset S$ denotes the subgroup of functions with $f(P_\infty) = 1$ for all real infinite primes then the torsion ideal is isomorphic to $F^+/F^{**} \times S_T \subset \mathrm{Sym}(F)$.

2. The boundary operator

We shall analyse the exact sequence

$$0 \to W(O_F) \to W(F) \xrightarrow{\ \partial\ } W(F/O_F)\ .$$

There is the decomposition

$$W(F/O_F) \simeq \sum_{\substack{P \\ \text{finite}}} W(F/O_F(P))\ .$$

and thus we may introduce

$$\partial_P: W(F) \to W(F/O_F(P)) .$$

By choosing a local uniformizer $\pi \in O_F(P)$ we can define

$$O_F(P) \to F/O_F(P)$$

by

$$\lambda \to (\lambda/\pi)$$

to yield the embedding of the residue field

$$F_P \to F/O_F(P) .$$

This will produce an isomorphism

$$W(F_P) \simeq W(F/O_F(P)) .$$

At a non-dyadic prime this does depend on the choice of local uniformizer, however, the restriction of this isomorphism to the fundamental ideal $J_P \subset W(F_P)$ is independent of the choice of local uniformizer.

It is also obvious that for each finite prime the composition

$$W(F) \xrightarrow{\partial_P} W(F_P) \xrightarrow{rk} Z/2Z$$

is independent of local uniformizer. This composition is computed as follows.

(2.1) Lemma: For $X \in W(F)$

$$rk\ \partial_P(X) = ord_P(dis\ X) \in Z/2Z .$$

Proof:

Since $X \to ord_P(dis\ X)$ (mod 2) is an additive homomorphism it is sufficient to check the assertion on rank 1 innerproducts. If $x \in F^*$ we write $x = \pi^k u$,

$u \in O_F(P)^*$. If k is even, $\partial_P(\langle x\rangle) = \partial_P\langle u\rangle = 0$. If k is odd then $\partial_P\langle x\rangle = \partial_P\langle \pi u\rangle = \langle u\rangle_P \in W(F_P)$. Here $\langle u\rangle_P$ is the Witt class of the rank 1 form on F_P defined by the image of u in the residue field. ■

If P is dyadic then $\mathrm{rk}: W(F_P) \simeq Z/2Z$ so we have determined ∂_P for dyadic primes.

We also know that for non-dyadic primes

$$\partial_P: J^2 \to J_P \subset W(F_P) .$$

We wish to expand on this observation

(2.2) <u>Lemma</u>: <u>For</u> X, Y <u>in</u> $W(F)$

$$\partial_P(XY) = \partial_P(X)\partial_P(\langle\pi\rangle Y)) + \partial_P(\langle\pi\rangle X)\partial_P(Y) .$$

<u>Proof</u>:

Again we need only check this on rank one forms. For x, y in F^* we write $x = \pi^k u$, $y = \pi^j v$ so that $xy = \pi^{k+j}uv$. Check the assertion in this case and it follows in general. ■

Suppose $X \in J$, then $\operatorname{dis} X = \operatorname{dis}(\langle\pi\rangle X)$. Suppose $Y \in J$ also. Set $\varepsilon_1 = \mathrm{rk}(\partial_P(X)) = \mathrm{ord}_P(\operatorname{dis} X) \in Z/2Z$ and $\varepsilon_2 = \mathrm{rk}(\partial_P(Y)) = \mathrm{ord}_P(\operatorname{dis} Y) \in Z/2Z$.

(2.4) <u>Lemma</u>: <u>For</u> X, Y <u>in</u> J $\operatorname{dis}(\partial_P(XY)) = (-1)^{\varepsilon_1\varepsilon_2}(\operatorname{dis}\partial_P(Y)\operatorname{dis}\partial_P(\langle\pi\rangle Y))^{\varepsilon_1} \times (\operatorname{dis}\partial_P(X)\operatorname{dis}\partial_P(\langle\pi\rangle X))^{\varepsilon_2}$.

Proof:

$$\begin{aligned}
\mathrm{dis}(\partial_P(XY)) &= \mathrm{dis}(\partial_P(\langle\pi\rangle X)\partial_P(Y) + \partial_P(X)\partial_P(\langle\pi\rangle Y)) \\
&= (-1)^{\varepsilon_1\varepsilon_2}(\mathrm{dis}(\partial_P(\langle\pi\rangle X)\partial_P(Y)) \times \mathrm{dis}(\partial_P(X)\partial_P(\langle\pi\rangle Y)) \\
&= (-1)^{\varepsilon_1\varepsilon_2}(\mathrm{dis}\ \partial_P(Y))^{\varepsilon_1}(\mathrm{dis}(\partial_P(\langle\pi\rangle X))^{\varepsilon_2} \\
&\quad \times (\mathrm{dis}\ \partial_P(Y))^{\varepsilon_1}(\mathrm{dis}(\partial_P(\langle\pi\rangle X))^{\varepsilon_2} \\
&= (-1)^{\varepsilon_1\varepsilon_2}(\mathrm{dis}\ \partial_P(Y)\mathrm{dis}\ \partial_P(\langle\pi\rangle Y))^{\varepsilon_1} \times (\mathrm{dis}\ \partial_P(X)\mathrm{dis}\ \partial_P(\langle\pi\rangle X))^{\varepsilon_2}. \quad \blacksquare
\end{aligned}$$

We shall use this to show

(2.5) Lemma: For any finite prime

$$\partial_P | J^3 \to W(F_P)$$

is trivial.

Proof:

We need only consider A, B, C in J with $\mathrm{rk}\ \partial_P(A) = r_1$, $\mathrm{rk}\ \partial_P(B) = r_2$, and $\mathrm{rk}\ \partial_P(C) = r_3$. Take $X = A$ so that $\varepsilon_1 = r_1$ and $Y = BC$. Note that $BC \in J^2$ so that $\mathrm{dis}(BC) = 1$ and $\mathrm{rk}\ \partial_P(BC) = 0 = \varepsilon_2$. Then $\mathrm{dis}\ \partial_P(ABC) = (\mathrm{dis}\ \partial_P(BC)\mathrm{dis}(\langle\pi\rangle BC))^{r_1}$. Now $(\mathrm{dis}\ \partial_P(BC))^{r_1} = (-1)^{r_1r_2r_3}(\mathrm{dis}\ \partial_P(C)\mathrm{dis}\ \partial_P(\langle\pi\rangle C))^{r_1r_2} \times (\mathrm{dis}\ \partial_P(B)\mathrm{dis}(\partial_P(\langle\pi\rangle B))^{r_1r_3}$ while $(\mathrm{dis}\ \partial_P(\langle\pi\rangle BC))^{r_1} = (-1)^{r_1r_2r_3}(\mathrm{dis}\ \partial_P(C)\mathrm{dis}\ \partial_P(\langle\pi\rangle C))^{r_1r_2} \times$ $(\mathrm{dis}\ \partial_P(\langle\pi\rangle B)\mathrm{dis}\ \partial_P(B))^{r_1r_3}$. We now see that

$$\mathrm{dis}\ \partial_P(ABC) = 1 \in F_P^*/F_P^{**}$$

and since $\partial_P(ABC) \in J_P$ we have $\partial_P(ABC) = 0$. Any element in J^3 is a sum of such triple products. ■

This proves that J^3 lies in the image of $W(O_F) \to W(F)$. More of this later. We have shown also that for each non-dyadic prime there is a homomorphism $\partial_P: J^2/J^3 \to J_P \subset W(F_P)$ which is independent of the choice of local uniformizer.

(2.6) [M-H, p. 96]: <u>If</u> P <u>is a finite non-dyadic prime and</u> $X \in J^2$ <u>then</u>

$$\partial_P(X) = c(X)(P) .$$

<u>Proof</u>:

The assertion means that $\text{dis}\ \partial_P(X) = 1$ if and only if $c(X)(P) = 1$. Using the realization for Hilbert symbols any element of J^2 can be written as $X = (\langle x\rangle - \langle 1\rangle)(\langle y\rangle - \langle 1\rangle) + Y$ where $Y \in J^3$. Therefore

$$c(X)(P) = ((x,y))(P)$$
$$\partial_P(X) = \partial_P(\langle xy\rangle - \langle x\rangle - \langle y\rangle + \langle 1\rangle) .$$

The computation of Hilbert symbols at non-dyadic prime is, as pointed out in [O'M, p. 166], quite easy. In terms of a local uniformizer π in $O_F(P)$ write $x = \pi^j u$, $y = \pi^k v$. If $j \equiv k \equiv 0 \pmod 2$ then $\partial_P(\langle xy\rangle - \langle x\rangle - \langle y\rangle + \langle 1\rangle) = 0$ but $((x,y))(P) = ((u,v))(P) = 1$. If $j \equiv 1 \pmod 2$ and $k \equiv 0 \pmod 2$ then

$$\text{dis}\ \partial_P(\langle xy\rangle - \langle x\rangle - \langle y\rangle + \langle 1\rangle) = v \in F_P^*/F_P^{**}$$

while

$$((x,y))(P) = ((\pi,v))(P) = 1$$

if and only if v is a square in F_P^*. Finally when $j \equiv k \equiv 1 \pmod 2$ $\text{dis}(\partial_P(\langle xy\rangle - \langle x\rangle - \langle y\rangle + \langle 1\rangle) = -uv$ in F_P^*/F_P^{**}. This time

$$((x,y))(P) = ((\pi u, \pi v))(P) = ((-uv, \pi))(P) = 1$$

if and only if $-uv$ is a square in the residue field. ■

We can use this to characterize the cokernel of

$$\partial\colon W(F) \to W(F/O_F)\ .$$

We write

$$W(F/O_F) = \sum_{\substack{P \\ \text{finite}}} W(F_P)\ .$$

In the sum on the right there is the subgroup $\sum_{\substack{P \\ \text{finite non-dyadic}}} J_P$. We claim

$$\partial\colon J^2 \to \sum_{\substack{P \\ \text{non-dyadic}}} J_P$$

is an epimorphism. There is always a dyadic prime in F so using that fact and the realization of Hilbert symbols we can find a pair x,y in F^* with $((x,y))(P)$ arbitrarily prescribed at the finite non-dyadic primes. We use $\partial((\langle x\rangle-\langle 1\rangle)(\langle y\rangle-\langle 1\rangle))$ and (V.2.6). Thus the issue becomes that of prescribing $\mathrm{ord}_P(x) \pmod 2$ at all finite primes for $x \in F^*/F^{**}$. Let $\tilde{M}_F \subset M_F$ be the subset of finite primes. A prescription then is a function

$$\varepsilon\colon \tilde{M}_F \to Z/2Z$$

where $\varepsilon(P) = 0$ for almost all finite primes. The prescriptions form an additive abelian group. There is a natural epimorphism of this group onto the quotient group $C(F)/C(F)^2$. Namely to each prescription associate

$$\prod_{P \text{ finite}} |P|^{\varepsilon(P)}$$

in $C(F)/C(F)^2$. There results an exact sequence

$$1 \to F^{ev}/F^{**} \to F^{*}/F^{**} \to \{\varepsilon\colon \tilde{M}_F \to Z/2Z\} \to C(F)/C(F)^2 \to 1 .$$

Now we have established

(2.7) [M-H, p. 94]: <u>The cokernel of</u> $\partial\colon W(F) \to W(F/O_F)$ <u>is isomorphic to</u> $C(F)/C(F)^2$. ■

This should be compared with the results in (IV.7). The subgroup $F^{ev} \subset F^{*}$ consists of those elements which have even order at every finite prime. This group will figure heavily in the next section.

3. The ring $W(O_F)$

We shall begin with the characterization of the kernel of ∂.

(3.1) <u>Lemma</u>: <u>If</u> $X \in W(F)$ <u>then</u> $\partial(X) = 0 \in W(F/O_F)$ <u>if and only if</u>

1) $\mathrm{ord}_P(\mathrm{dis}\ X) \equiv 0 \pmod 2$, <u>all finite primes</u>
2) $c(X)(P) = 1$, <u>all non-dyadic finite primes</u>.

<u>Proof</u>:

Obviously $\langle 1 \rangle$ lies in the kernel of ∂ so we need only consider $X \in J$. By (V.2.1) we find $\mathrm{rk}\ \partial_P(X) = 0$ in F_2 if and only if $\mathrm{ord}_P(\mathrm{dis}\ X) \equiv 0 \pmod 2$. Then write $X = \langle \mathrm{dis}\ X\rangle - \langle 1\rangle + Y$ for $Y \in J^2$. From the definition of boundary, $\partial(\langle \mathrm{dis}\ X\rangle - \langle 1\rangle) = 0$ if and only if $\mathrm{dis}\ X \in F^{ev}/F^{**}$. Since $Y \in J^2$ we know by (V.2.6) that $\partial(Y) = 0$ if and only if $c(Y)(P) = 1$ at all finite non-dyadic primes. Finally, $c(X) = ((\mathrm{dis}\ X, -1))c(Y)$ and hence for non-dyadic finite primes

$$c(X)(P) = c(Y)(P)$$

if $\mathrm{ord}_P(\mathrm{dis}\ X) \equiv 0 \pmod 2$. ■

This allows us to introduce

$$\mathrm{Sym}(O_F) = F_2 \times F^{ev}/F^{**} \times \tilde{S}$$

as a subring of $\mathrm{Sym}(F)$. By $\tilde{S} \subset S$ we mean the subgroup of $f\colon M_F \to Z^*$ such that $f(P) = 1$ for all non-dyadic finite primes. We have an epimorphism

$$W(O_F) \to \mathrm{Sym}(O_F) \to 1$$

From (V.3.1) it follows immediately that $J^3 \subset W(F)$ lies in the image of $W(O_F) \to W(F)$. Let us denote by $J_3 \subset W(O_F)$ the ideal corresponding to J^3. We can state

(3.2) Theorem: If F is formally non-real then

$$W(O_F) \simeq \mathrm{Sym}(O_F)$$

while if F is formally real

$$0 \to J_3 \to W(O_F) \to \mathrm{Sym}(O_F) \to 0$$

and

$$\mathrm{Sgn}\colon J_3 \simeq 8Z(F) \;. \;\blacksquare$$

We shall discuss the group F^{ev}/F^{**} which is the analogue of $\mathrm{Gen}(E/F)$. If a unit in O_F^* is the square of an element in F^* then it is the square of a unit. For that reason there is an embedding

$$1 \to O_F^*/O_F^{**} \to F^{ev}/F^{**}$$

We can explain the cokernel as follows. Let $\tilde{C}(F) \subset C(F)$ be the subgroup of ideal classes of order ≤ 2. If $x \in F^{ev}$ then there is a fractional O_F-ideal A with $A^2 = xO_F$. Send x to $|A| \in \tilde{C}(F)$. If $B^2 = xO_F$ also then $(AB^{-1})^2 = O_F$ so that $A = B$. If x is replaced by xy^2 then A is replaced by yA. Thus we have a well defined epimorphism

$$F^{ev}/F^{**} \to \tilde{C}(F) \to 1 .$$

If $A^2 = xO_F$ and $A = zO_F$ then for some unit $u \in O_F^*$

$$uz^2 = x$$

Thus we say

(3.3) <u>Lemma</u>: <u>There is a short exact sequence</u>

$$1 \to O_F^*/O_F^{**} \to F^{ev}/F^{**} \to \tilde{C}(F) \to 1 . \blacksquare$$

Actually there is an embedding of F^{ev}/F^{**} into $W(O_F)^*$. For $x \in F^{ev}$ write $A^2 = xO_F$ and on A introduce

$$b(a_1, a_2) = a_1 a_2 / x .$$

This is a rank 1 innerproduct space over O_F since $A^{-1} \simeq \mathrm{Hom}_{O_F}(A, O_F)$ and $x \to \langle A, b \rangle$ gives the embedding. We must say, however, that we have no idea as to whether or not a diagonalization theorem analogous to (IV.6.1) is valid for $W(O_F)$. Using the discriminant there is an epimorphism

$$W(O_F) \to \tilde{C}(F)$$

and we leave to the reader

(3.4) <u>Exercise</u>: <u>The following are equivalent for</u> $X \in W(O_F)$.

1) X <u>is represented by a symmetric innerproduct on a finitely generated free</u> O_F<u>-module</u>.

2) $\mathrm{dis}\, X \in F^{ev}/F^{**}$ <u>is a unit from</u> O_F^*/O_F^{**}.

3) X lies in the kernel of

$$W(O_F) \to \tilde{C}(F) \to 1 . \blacksquare$$

Let us discuss the ring $\mathrm{Sym}(O_F)$ which is finite. First $\tilde{S}$ has order 2^{r_1+d-1} where d is the number of dyadic primes in F and r_1 is the number of real infinite primes. The Dirichlet Units Theorem can be used to compute the order of O_F^*/O_F^{**}. It has order $2^{r_1+r_2}$ where r_2 is the number of complex infinite primes in F. Therefore

(3.5) Lemma: The order of $\mathrm{Sym}(O_F)$ is equal to

$$\#(\tilde{C}(F))\cdot 2^{2r_1+r_2+d} . \blacksquare$$

If F is formally non-real then r_1 is 0 and $W(O_F)$ has order

$$\#(\tilde{C}(F))\cdot 2^{r_2+d}$$

The additive order of $\langle 1\rangle$ in $W(O_F)$ will be 2,4 or 8 corresponding to the Stufe values 1,2 and 4.

For example let $F = Q(\sqrt{m})$ where m is a negative square free integer. Use (I.6.10) for the extension $Q(\sqrt{m})/Q$ to see that $\tilde{C}(F)$ and the cohomology group $H^1(C_2;C(F))$ have the same order. Then from (I.4.4) the order of $\tilde{C}(F)$ is 2^{N-2} where N is the total number of primes, including the real infinite prime, which ramify. If m is even 2 ramifies. If m is odd then 2 ramifies for $m \equiv 3 \pmod 4$, is inert for $m \equiv 5 \pmod 8$ and splits for $m \equiv 1 \pmod 8$. Thus $d = 1$ unless $m \equiv 1 \pmod 8$ in which case it is 2. Of course r_2 is always 1.

Now $Q(\sqrt{m})$ contains the square root of -1 only for $m = -1$. If $m < -1$ and $m \not\equiv 1 \pmod 8$ then $((-1,-1)) = 1$. This is clear since $Q(\sqrt{m})$ contains only one dyadic and no real infinite primes. We know $((-1,-1))(P) = 1$ at all non-dyadic primes so $((-1,-1)) = 1$ by reciprocity and $\langle 1\rangle \in W(O_F)$ has additive order 4 in these cases. For $m \equiv 1 \pmod 8$ there are two dyadic primes P_1, P_2 and in fact $((-1,-1))(P_1) = ((-1,-1))(P_2) = -1$. Simply note that the localized completion $\tilde{F}(P_i)$, $i = 1,2$, is an extension of degree 1 of the dyadic rationals $\tilde{Q}(2)$. Thus $((-1,-1))(P_i) = (-1,-1)_2 = -1$. Hence $\langle 1\rangle$ has order 8 if $m \equiv 1 \pmod 8$.

If F is formally real then the torsion ideal in $W(O_F)$ is also the nil-radical in this ring and embeds in $\mathrm{Sym}(O_F)$. Using (V.1.9) it follows that the torsion ideal is isomorphic to

$$F_+^{ev}/F^{**} \times \tilde{S}_T$$

where $F_+^{ev} \subset F^{ev}$ is the subgroup of elements positive with respect to every ordering of F and $\tilde{S}_T \subset \tilde{S}$ is the subgroup of elements such that $f(P_\infty) = 1$ for all real infinite primes.

The order of $\tilde{S}_T$ is 2^{d-1}. We can estimate the order of F_+^{ev}/F^{**}. Define

$$F^{ev}/F^{**} \to (Z^*)^{r_1} ,$$

a homomorphism which to $x \in F^{ev}/F^{**}$ assigns a sequence of 1 and -1 depending on the sign of x with respect to the various orderings. The kernel of this homomorphism is F_+^{ev}/F^{**}, therefore we can say

(3.6) <u>Lemma</u>: <u>The order of the torsion ideal in</u> $W(O_F)$ <u>is at least equal to</u>

$$\#(\tilde{C}(F))2^{r_2+d-1} . \blacksquare$$

As we mentioned earlier there is

$$1 \to F^{ev}/F^{**} \to W(O_F)^* .$$

We would like to know when the image generates $W(O_F)$ additively. The only answer of which we are aware is not at all effective, but it does raise some interesting problems. Think of F^{ev}/F^{**} as a vector space over F_2. Form the symmetric product $(F^{ev}/F^{**}) \circ (F^{ev}/F^{**})$. The correspondence $x \circ y \to ((x,y)) \in \tilde{S}$ yields a homomorphism

$$(F^{ev}/F^{**}) \circ (F^{ev}/F^{**}) \to \tilde{S} .$$

This is an epimorphism if and only if the image of $F^{ev}/F^{**} \to W(O_F)^*$ generates $W(O_F)$ additively. If F is formally non-real and contains exactly one non-dyadic prime then $\tilde{S}$ is trivial and we obtain diagonalization by default. If $W(O_F)$ is torsion free then diagonalization is also valid [M-H, p. 95]. Let us take up some elementary examples for which diagonalization will fail. Suppose $p > 0$ is an odd prime, then put $F = Q(\sqrt{p})$. The class number of F is odd. If $p \equiv 1 \pmod 4$ use (I.6.11). For $p \equiv 3 \pmod 4$ use (I.4.4, I.6.10) on $Q(\sqrt{p})/Q$ noting that -1 is not a norm from $Q(\sqrt{p})$ and the Gen group is cyclic of order 2 implying $H^1(C_2;C(F)) = \{1\}$. This means $F^{ev}/F^{**} = O_F^*/O_F^{**}$. As an F_2 basis use -1 and a fundamental unit, U. The symmetric product of O_F^*/O_F^{**} has basis $-1 \circ -1$, $U \circ U$ and $-1 \circ U$ so that the image of the symmetric product in $\tilde{S}$ is spanned by $((-1,-1))$, $((U,U))$ and $((-1,U)) = ((-U^2,U)) = ((-U,U))((U,U)) = ((U,U))$. We are in immediate trouble if $p \equiv 1 \pmod 8$. In that case F has two dyadic primes and two orderings so that $\tilde{S}$ has dimension 3 over F_2. We are just as bad off when $p \equiv 3 \pmod 4$. There is no loss in assuming U is positive in at least one of the two orderings but since $U\bar{U} = 1$ we know U is positive in both orderings and so $((U,U)) = 1$. However $\dim_{F_2} \tilde{S} = 2$, thus we are done. The case $p \equiv 5 \pmod 8$ is the one for which $W(O_F)$ has no torsion and therefore diagonalization up to Witt is valid.

It appears to us that this too is a contrast between $W(O_F)$ and $H(O_E)$. We have not seen any pattern that would produce a general result.

References

[ACH] J.P. Alexander, P.E. Conner and G.C. Hamrick, Odd Order Group Actions and Witt Classification of Innerproducts, Springer-Verlag, Lecture Notes in Mathematics No. 625 (1977).

[B] N. Bourbaki, Commutative Algebra, Hermann, Addison-Wesley (1972).

[C] H. Cohn, A Second Course in Number Theory, John Wiley and Sons (1962).

[La] W. Landherr, Aequivalenz Hermitescher Formen ueber einen beliebigen algebraishen Zahlkoeper, Abh. Math. Sem. Hamburg Univ. vol II, p. 245 (1935).

[L] S. Lang, Algebraic Number Theory, Addison-Wesley Series in Mathematics (1970).

[M-H] J.W. Milnor and Dale Husemoller, Symmetric Bilinear Forms, Springer-Verlag, Ergebnisse Series, Bd 73 (1973).

[O'M] O.T.O'Meara, Introduction to Quadratic Forms, Springer-Verlag, Grundlehren der Math. Wiss., vol. 117 (1963).

[St] N.W. Stoltzfus, Unraveling the integral knot concordance group, Memoirs Amer. Math. Soc., vol. 12, issue 1, number 192 (1977).

[We] E. Weiss, Algebraic Number Theory, McGraw-Hill, International Series in Pure and Applied Math. (1963).

Symbol List

E, F, fields

E/F, E is a quadratic extension of F

O_E, O_F, rings of algebraic integers

$\mathcal{P}$, P, finite primes in O_E, O_F

$\mathcal{P}_\infty$, P_∞, infinite primes in E, F

$O_E(\mathcal{P})$, $O_F(P)$, local integers

π, local uniformizer, $\mathrm{ord}_{\mathcal{P}}(\pi) = 1$ or $\mathrm{ord}_P(\pi) = 1$

$\mathrm{ord}_{\mathcal{P}}(\cdot)$, $\mathrm{ord}_P(\cdot)$, $\nu_{\mathcal{P}}(\cdot)$, $\nu_P(\cdot)$, order at a finite prime

$m(\mathcal{P})$, $m(P)$, maximal ideals in $O_E(\mathcal{P})$, $O_F(P)$

$\tilde{E}(\mathcal{P})$, $\tilde{F}(P)$, $\tilde{O}_E(\mathcal{P})$, $\tilde{O}_F(P)$, localized completions

$E_{\mathcal{P}}$, F_P, residue fields

$H^*(C_2;O_E(\mathcal{P})^*)$, cohomology of C_2 acting by conjugation

$\mathcal{D}_{E/F}$, different ideal of the extension E/F

$H^2(C_2;E^*)$, F^* modulo norms from E^*

$(y,\sigma)_P$, a Hilbert symbol (I.2)

C(E), C(F), ideal class groups

K(E/F), kernel of $C(F) \to C(E)$

$|A|$, ideal class of the fractional ideal A

Gen(E/F), a group in (I.3)

Iso(E/F), a group in (I.4)

I(E), I(F), groups of fractional ideals

Sc, a homomorphism in (I.5) also [St, sec. 4]

H(E), Witt ring of Hermitian innerproducts (II)

$W^{\perp}$, orthogonal subspace (II.1)

(V,h), Hermitian innerproduct space over E

rk, rank mod 2 (II.3)

J, fundamental ideal (II.3)

Gp(E), a group (II.3.1)

W(E), Witt ring of symmetric innerproducts over E

Z(E), a direct sum of Z with itself

Sgn: $H(E) \to Z(E)$, total signature (II.4)

M, torsion O_E-module

$M_{\mathcal{P}}$, $\mathcal{P}$-primary torsion submodule of M

E/K, quotient of E by a fractional ideal $K = \overline{K}$

(M,b), torsion u-innerproduct with values in E/K

$H_u(E/K)$, Witt group of torsion u-innerproducts

$K_{(\mathcal{P})}$, localization of K (III.2)

$E/K_{(\mathcal{P})}$, quotient of E by the localization

$\Gamma_{\mathcal{P}} \subset E/K_{(\mathcal{P})}$, 1-dimensional subspace over $E_{\mathcal{P}}$ (III.3)

ρ, localizer (III.3.10), (IV.4.3)

$\mathfrak{D}^{-1}$, inverse different of E over Q (III.5)

$tr_{E/Q}$, trace of E/Q

$tr_{E_{\mathcal{P}}/F_p}$, trace of the residue field over its prime field

∂, boundary in the Knebusch sequence (IV.3)

I(u,K), ideal in $H(O_E)$ isomorphic to $H_u(K)$ (IV.6)

(P,h), a u-innerproduct space over K with P projective (IV)

$H(O_E)$, Witt ring of Hermitian innerproducts (IV.6)

ω, Artin map, (I.5)

dis, discriminant, (II.3)

type I, type II, types of ramified primes (I.1.4)

type (α), type (β), another classification of ramified (III.3.13)

$c(\cdot)$, Hasse-Minkowski invariant (V.1)

S, group of symbols with reciprocity (V.1)

$Sym(\cdot)$, ring analogous to $Gp(\cdot)$, (V.1)

F^{ev}, group of elements in F^* with even order at all finite primes (V.3)

$\tilde{C}(\cdot)$, subgroup of elements in the ideal class group with order ≤ 2

Printed and bound by CPI Group (UK) Ltd, Croydon, CR0 4YY

27/10/2024

14580156-0001